TOWARD
CREATIVE SYSTEMS
DESIGN

TOWARD CREATIVE SYSTEMS DESIGN

HENRY C. LUCAS, JR.

1974

COLUMBIA UNIVERSITY PRESS
New York and London

Henry C. Lucas, Jr., is an associate professor in the Graduate School of Business Administration, New York University.

Library of Congress Cataloging in Publication Data

Lucas, Henry C
  Toward creative systems design.

  Bibliography: p.
  1. Management information systems. 2. Systems engineering. I. Title.
T58.6.L82   658.4'03   74–4129
ISBN 0–231–03791–0

TO
MY
PARENTS

PREFACE

DURING THE PAST DECADE many computer-based information systems have been developed in private and public sector organizations. These systems have a tremendous potential for improving organizational performance and decisionmaking and reducing some of the drudgery of information processing. Unfortunately, research experiences in a broad sample of organizations indicate that achievement of this potential has not even been approached. In fact, some of the systems being developed may actually be dysfunctional to the organization.

This research has demonstrated that many computer-based information systems today are mundane and uninteresting. Chapter 1 of this monograph examines some development trends and predicts that future efforts will be directed toward more sophisticated applications. However, if success has not really been attained in designing simple systems, how can it be expected in the design of more complex systems? A concern over the failure of many of our information systems stimulated the preparation of this monograph on creative systems design.

The purpose of the monograph is to coalesce scattered research results and develop recommended techniques for systems design which consciously consider organizational behavior problems. The monograph is intended for students in systems design courses and for computer professionals in systems design and management positions. No particular background on the part of the reader is assumed; however, some actual systems design experience or classroom exposure to design exercises should make the monograph more meaningful to the reader.

The title, *Toward Creative Systems Design,* was chosen because in general the techniques advocated here are new approaches to design problems. Some of the individual techniques recommended may have been used before, but rarely has a design strategy been created to apply consciously a combination of these techniques to overcoming organizational behavior problems in design. Given the current level of knowledge, there are no formulas or rules, and creativity is required to develop an approach to systems design using techniques appropriate for each situation.

Part I stresses that organizational behavior problems are the most crucial ones in systems design. The major problems are discussed in light of this view. The central philosophy of creative systems design is that users, rather than systems analysts, should design information systems. Creative systems design means paying special attention to change processes, the physical user interface, and the quality of a system as defined by the user. These topics are discussed in Part II. If the technical components of a system do not work properly or the design project is not managed successfully, creative design cannot succeed. Therefore, mechanical aids to systems design and project management are treated in Part III. (This monograph stresses organizational behavior problems in design. Some technical topics are considered, but in general technical issues are not covered comprehensively unless they have an important impact on organizational behavior problems.)

I acknowledge the contribution of several individuals to the book. Rodney Plimpton and Tilmon Kreiling were responsible for much of the work on two of the examples presented in Chapter 6. Mrs. Lynne Rosenthal has had the patience to manage the typing of revisions to the manuscript while contending with other heavy demands on her time. Finally, my wife Ellen managed to provide invaluable editorial and spiritual support while having and caring for our first child and simultaneously completing the requirements for a Master's degree. She has my continuing admiration and gratitude for her accomplishments and contribution to the monograph.

January 1974 *Henry C. Lucas, Jr.*

CONTENTS

PART ONE

INTRODUCTION

CHAPTER ONE

INTRODUCTION AND MOTIVATION
FOR CREATIVE SYSTEMS DESIGN

INFORMATION SYSTEMS

THE DESIGN OF information systems is one of the most challenging activities in an organization today. In fifteen years or less, the information services department has become a major component of the organization, responsible for a large number of professional employees and a huge investment in equipment and software. Such phenomenal growth is particularly impressive given the fact that real industrial interest in information systems began in the late 1950s and academic recognition of the field did not occur until the 1960s. Unfortunately, the rapid growth of information systems has resulted in numerous problems, particularly in systems design. Many of the problems are quite serious and our ability to solve them will determine whether or not information systems achieve their potential. This book is devoted to a discussion of various creative solutions to these problems.

As a part of a continuing research program on the relationship between users, information systems, and the information services department staff, we have conducted surveys of nearly 2000 people in over 15 companies. Each of the questionnaires asked for general comments at the end of the structured questions. An analysis of these user reactions indicates that there are crucial problems in the design and operation of information systems. All the difficulties discussed below have been mentioned at least once on questionnaires and in interviews, and most of the problems have been mentioned repeatedly. In the following discussion we do not want to suggest that every organization has experienced these problems with all of its systems. However, these problems have occurred before and are likely to recur if preventive measures are not taken.

According to our research, users do not understand much of the output they receive; schedules are not kept and too much information is outdated when received. There is duplication of input and output, and changes are frequently made without consulting users. Attainment of accuracy is a constant problem; if there are a few errors, users often discount all the information provided by a system. Many users complain of information overload; massive amounts of data are provided which cannot be digested by the decisionmaker. There are also many complaints about the cost and development time for new applications. Changes in existing systems are difficult to arrange. For many users, computer systems are considered unresponsive, inflexible, and not worth the effort and expense to develop.

In this monograph, we examine three major problem areas in systems design; the technical aspects of design, organizational behavior, and project management. The philosophy of creative systems design is focused on solutions to organizational behavior problems which have largely been ignored in the design of present systems. In addition to the need for new approaches to systems design evidenced by the user reactions described above, an examination of the current state of the art of information systems and forecasts of the characteristics of future systems also suggest strongly that better design methods are needed. However, before additional evidence supporting the development of a philosophy of creative design or the philosophy itself can be presented, it is necessary to define information systems and develop a conceptual basis for further discussions. The information systems field is ill-structured; several frameworks for information systems have been proposed. These frameworks provide conceptual models for discussion purposes; their descriptions follow a definition of information systems.

DEFINITION

An information system is designed to support decisionmaking (Lucas, 1973a). A system is a set of organized procedures which are executed to produce a desired output. Information, itself, is some tangible or intangible entity which reduces uncertainty about a future state or event. For example, receiving information that a trucking strike has been canceled reduces our uncertainty about tomorrow's delivery of a product.

DECISIONMAKING

The main focus of this definition for information systems is on decisionmaking; Simon has defined three major steps in the decisionmaking process (Simon, 1965). The first one is the Intelligence phase, which consists in searching the environment or becoming aware of a situation requiring a decision. The second phase is Design, which includes enumerating and evaluating the various possible alternatives. The final phase, Choice, is the selection of an alternative delineated during the Design phase. It is useful to add an additional step to those defined by Simon; the Implementation phase involves seeing that the decision is carried out.

An information system can play an important role in all parts of the decisionmaking process outlined above. During the Intelligence phase a system can provide signals that a decision needs to be made. In this sense it aids the decisionmaker in "problem finding" (Pounds, 1969). Information systems providing exception reports, such as data on stock-outs, are particularly important during the Intelligence phase.

During the Design phase the information system may present the alternatives. For example, it can supply the names of suppliers who sell the out-of-stock items. The information system may also evaluate the different alternatives: what are the costs of purchasing from the various suppliers, given transportation charges, quantity discounts, price breaks, etc.?

For the Choice phase, often a decision rule is specified which is automatically executed by the information system. During Implementation an information system may check to see that the decision has been executed. In the example above, the system could be designed to monitor the receipt of the item by a certain time, or to indicate that action is needed to expedite the shipment.

Clearly there are many types of information systems. Organizations had these systems long before the topic became a separate field of study and long before computers were available. Many recent systems use a computer for processing because of a high volume of transactions, the large amount of data stored on files, and the limited time to perform the many calculations required to process the data. Here we are concerned with the design of computer-based information systems. Although much of the discussion is relevant to other types of information systems, these issues are particularly important for information systems which use computers. Typically, computer-based information systems involve many members of the organization. A large investment is required to develop and operate these systems, and they are of crucial importance to the continued functioning of the organization.

FRAMEWORKS FOR INFORMATION SYSTEMS

A framework is a conceptual model which provides a perspective for viewing information systems. We shall develop a framework to provide a common basis for discussing information systems. Because information systems support decisionmaking, our framework will classify decisions into different categories. We shall use the framework to organize our thoughts about existing systems and systems design and to predict future trends in systems development. For viewing existing information systems and thinking about system design, we shall use a framework developed by Robert Anthony. An extension to the Anthony framework by Gorry and Scott Morton will be used to predict future trends in systems design.

THE ANTHONY FRAMEWORK

A framework for planning and control decisions developed by Anthony includes three types of decisions. The first is strategic planning, "the process of deciding on objectives of the organization, on changes in these objectives, on the resources used to obtain these objectives, and on the policies that are used to govern the acquisition, use, and disposition of these resources." (Anthony, 1965, p. 24). As one would expect, these decisions tend to be primarily associated with higher levels of management. They are concerned with long-range problems as opposed to day-to-day operations.

The next decision category distinguished by Anthony is managerial control, which deals primarily with financial and personnel problems. It is defined as "the process by which managers assure that resources are obtained and used effectively and efficiently in the accomplishment of the organization's objectives" (Anthony, 1965, p. 27). Finally, the last decision category is operational control, which is "the process of assuring that specific tasks are carried out effectively and efficiently" (Anthony, 1965, p. 69).

Operational control decisions have a major impact on the day-to-day operations of the organization. The bulk of the activities of lower level management fall into this category.

Given these three decisionmaking categories, what are the broad information requirements associated with each? In strategic planning, the decisionmaker has a need for a rather complex data, though it may be in aggregated and summary form. The data are not frequently updated, and often they are developed from an external source rather than from within the organization. Decisions of this type have a long-range horizon; for example, should we introduce product X? Decisions in the operational control category require detailed information, immediate updating, and greater accuracy; also the information is generally developed internally as a part of the operation of the organization. Managerial control decisions fall between strategic planning and operational control in their information requirements. See Table 1.1.

The decision categories and information requirements are viewed here as points on a continuum. For most decisions, exact

TABLE 1.1.   THE ANTHONY FRAMEWORK AND
INFORMATION CHARACTERISTICS

Normative Information Characteristics for:

Operational Control Decisions	Management Control Decisions	Strategic Planning Decisions
1. Very detailed data	Moderately detailed data	Aggregate data
2. Related to a specific task	Related to achievement of organization's objectives	Related to establishing broad policies
3. Frequently reported	Regularly reported	Infrequently reported
4. Historical data	Historical and predictive data	Predictive
5. Internally generated	Mostly internally generated	Externally generated
6. Very accurate	Accurate within decision bounds	Accurate in magnitude only
7. Repetitive	Exception reporting	Unique to problem under consideration
8. Often nonfinancial	Mainly financial	Often nonfinancial

classification will be difficult. However, we can see the usefulness of such a framework for communications and design purposes. We have observed systems where the vice president of finance received detailed reports on each item in inventory! For the types of decisions most frequently encountered by such a manager, this report was clearly irrelevant and inappropriate. The failure to have a descriptive model in mind when designing information systems results in much of the dissatisfaction with systems expressed by users.

DESCRIPTIVE VALIDITY

The Anthony framework has a great deal of intuitive appeal and certainly offers a simple, normative model which fits well with our definition of information systems as supporting decision-making. In addition to its normative implications, it is interesting to see how well the framework describes the current status of information systems. Has it been applied in systems design?

As a part of a larger study, seven manufacturing companies in the San Francisco Bay Area provided data on some of their existing computer-based information systems (see Lucas, 1974a). These companies make products ranging from foodstuffs to digital computers, and all use medium to large scale computers in their own operations. Twenty representative systems from a variety of applications areas were selected with the assistance of the manager of the information services department in each company. (Table 1.2 shows the comparability of the systems in the different

TABLE 1.2.   APPLICATION AREAS, COMPANIES, AND SYSTEMS

Application	Number of Companies	Number of Systems
Production control	4	4
Finance	3	3
Order processing	5	5
Payroll and personnel	5	6
Planning	1	2

companies.) A minimum of three companies were included for each general type of systems classification with the exception of planning; only one company had planning systems. Only major systems were studied; utility programs or trivial applications were not included.

The major unit of the analysis in this study was the number of reports, adopted as a reasonable surrogate for the number of decisions supported by each system. In batch systems, reports are often designed for only one function. Only about 150 of the total sample of 628 reports went to more than one decisionmaker; only 35 of these went to two or more decisionmakers who could potentially use the report for another decision. For an on-line system a single output serves many users, but there were only three such systems in the sample, representing less than 1% of the total number of reports.

Reports were gathered for all the systems, and the lead systems analyst was interviewed in depth by a researcher who coded the reports on decision type, purpose of the report (complexity), decisionmaker, frequency, the level of aggregation. The data represent only the systems analyst's perception of how the reports are being used. (A useful, but extensive, follow-up study would be to compare these perceptions with the actual use reported by decisionmakers.)

The data in Table 1.3 demonstrate the limited descriptive validity of the Anthony framework. Note that the decision type has been extended to include "none" and "execution" and that the managerial and strategic planning categories were combined. (The "none" and "execution" categories can best be described as "transactions processing" and will form the fourth category for a modified Anthony framework which we shall use for communications and design purposes. Thus our final framework categories are strategic planning, managerial control, operational control, and transactions processing.)

From the Anthony framework we predict that decisionmak-

TABLE 1.3. DECISION TYPE AND REPORT ACTIVITIES

Decision Type	Decision-maker	Reports	Purpose (Complexity)	Reports	Aggregation	Reports	Frequency	Reports
None *	Clerk †	51%	Product	47%	Full	81%	Daily	9%
	None	18%	Historical	44%	Summary	16%	Weekly	35%
					Exception	3%	Monthly	26%
							Other	30%
Execution	Clerk	27%	Historical	31%	Full	63%	Daily	9%
	Super clerk	20%	Maintenance	29%	Summary	20%	Weekly	33%
	Accountant	33%	Cost	29%	Exception	17%	Monthly	36%
							Other	16%
Operational	Clerk	28%	Historical	30%	Full	64%	Daily	26%
	Super clerk	31%	Maintenance	25%	Summary	9%	Weekly	34%
	Foreman/supervisor	30%	Problem/action	38%	Exception	27%	Monthly	19%
							Other	21%
Managerial-strategic	Foreman/supervisor	22%	Historical	17%	Full	49%	Daily	19%
	Management	66%	Problem/action	17%	Summary	42%	Weekly	32%
			Cost	26%	Exception	9%	Monthly	40%
			Planning/trend	37%			Other	9%
Spearman correlation with decision type	.33		.41		.12		-.11	
Significance	.001		.001		.001		.003	

All attributes showed significant differences for decision type with a χ^2 test, $P < .001$. (These results should be treated with caution, since some of the tables had more than 20% of their cells with expected frequencies of less than 5.)

* Even though there is no decision, the output was distributed.
† For example, 51% of the reports on which no decision is based are sent to clerks.

ing is positively associated with the type of decisions; that is, more strategic planning decisions are made by top management, and the data in Table 1.3 tend to support this prediction.[1] From the framework we also expect that the complexity of reports increases as the decision type moves from "none" to "managerial and strategic"; this prediction is also supported by the data in Table 1.3.

From a design standpoint, more exception reporting and more summaries should be associated with managerial and strategic decisions where there is less need for detailed data. The correlation here, though significant, is quite low. The framework also indicates that less frequent reporting occurs at the managerial and strategic levels. Again, the relationship is in the indicated direction, but the correlation is small.

Even with the limitations of the study, these results suggest that the Anthony framework does have descriptive validity, but there is certainly room to apply if further in the design of information systems.[2] It will be helpful for the reader to keep the modified framework in mind through the rest of the book when different problems and examples are discussed. How do the problems differ with the type of decision or decisionmaker involved? Are the suggested solutions equally applicable to transactional, operational control, managerial control, and strategic planning systems?

THE GORRY–SCOTT MORTON FRAMEWORK

A framework suggested by Gorry and Scott Morton is extremely useful in making predictions about the nature of fu-

[1] All results are significant at the .01 level or better; see Table 1.3.

[2] The discussion of information systems has focused on decisions and on applications. In using terms like "complexity" and "sophistication," we are referring to the decision and application setting, not the underlying computer technology. In fact, more sophisticated technology such as on-line systems is associated with more mundane and operational control systems according to the information characteristics of Table 1.3. For a further discussion of this point, see Lucas (1973a, especially Chapters 1 and 2).

ture information systems. This framework combines the work of Anthony with two decision categories derived from Simon's studies of decisionmaking.

Simon first describes programmed decisions, which are those of a routine and repetitive nature (Simon, 1965). For these decisions the organization has specific rules and procedures, and the decisions are often referred to as "structured." Nonprogrammed or "unstructured" decisions are those novel and infrequent "one-shot" situations which arise periodically. These two categories should actually be viewed as a continuum because it is difficult to make a precise classification of each decision. Though these categories are a useful way to view some decisions, they are not to much help in classifying the type of information required for decisions. However, when Simon's work is combined with Anthony's framework, the result is a very powerful descriptive model of information systems.

Gorry and Scott Morton used the Simon and Anthony frameworks to form a two-dimensional matrix (Gorry and Scott Morton, 1971). See Figure 1.1. The columns in this figure show Anthony's decision categories of operational control, managerial control, and strategic planning. The rows extend Simon's classification to

FIGURE 1.1.   THE GORRY AND SCOTT MORTON FRAMEWORK.

	Operational Control	Managerial Control	Strategic Planning
Structured	Accounts Receivable Order Entry	Budgeting Short-Term Forecasting	Tanker Fleet Mix Warehouse Location
Semi-structured	Inventory Control Production Scheduling	Variance Analysis	New Product Introduction
Unstructured	Cash Management	Personnel Management	R&D Planning

include structured, semistructured, and unstructured decisions. This framework is useful for classifying existing applications according to their contribution to decisionmaking. It serves to combine decision-oriented approaches with other applications-oriented frameworks. The Gorry-Scott Morton framework is very helpful in suggesting new directions for the design of computer-based information systems. Most applications today fall into the structured operational control cell of the framework, for example, accounts receivable, payroll, and inventory control.[3] However, as more of these systems are implemented, the trend for new applications will be to occupy cells in the lower right-hand side of Figure 1.1. That is, computer-based information systems will attack more managerial control and strategic planning decisions and less-structured problems in general.

With this framework, each axis represents a continuum and it is difficult to make rigid classifications. As we move into the semistructured and unstructured areas, the line separating these two categories from structured applications will also tend to move. Some authors maintain that, once an information system is developed, the decisions involved can no longer be classified as anything other than structured. While such a fine distinction is relatively unimportant, the trend that it signifies is of interest. Computer-based information systems should provide more structured information in what have been previously semistructured or unstructured decision situations. In these instances the computer-based system will probably not actually make the decision, but instead will provide different types of information in an organized form allowing the decisionmaker to develop new insights into his problem.

[3] These systems were undoubtedly chosen for implementation because they are the most straightforward and show the greatest cost saving. However, they are not the systems with the greatest potential for supporting decisionmaking.

FURTHER MOTIVATION FOR CREATIVE DESIGN

Earlier in this chapter we discussed user reactions to information systems and found that a number of serious problems exist which reduce the effectiveness of many systems. The conceptual base provided by our modified Anthony framework and the predictions for future trends in systems from the Gorry and Scott Morton framework provide additional motivation for the adoption of creative systems design techniques.

CURRENT STATE OF THE ART

Gorry and Scott Morton (1971) speculate that most information systems have been rather mundane in nature. Structured, operational control decisions have been automated first in most organizations, and they predict that new systems will attack unstructured problems. If users are dissatisfied with present systems and these systems are relatively simple, the implementation of more sophisticated systems is not likely to be successful. Is there any evidence which supports the claim that existing systems have in general attacked unsophisticated decisions? Two empirical studies are described below which support the contention that information systems have had a limited direct impact on management and decisionmaking.

A study was conducted by Churchill in which managers, users, and computer management were interviewed in detail in a number of companies (Churchill et al., 1969). The researchers concluded that the literature presents a far more advanced picture of computer usage in business than what actually exists. They indicated that the computer has achieved much in clerical operations, but has been less frequently applied in other areas. There is some move toward systems integration, but there are still many diverse applications.

There was also a trend toward delegating more decisions to computers, but there had been little or no impact of computer-based information systems on the higher levels of management. As one might expect, computers were formalizing the organization's information systems. New applications seemed to be broad in scope, employing computer systems as more than a book-keeping tool. These new applications also tended to cross functional boundaries.

The information services department within an organization has two different functions. First, it operates a production shop running systems and, second, it is a research and development organization developing new systems. Unfortunately, management of the computer resource was characterized by isolation; it was left to nonmanagement, technically oriented people. Finally, the authors concluded that "generally applicable measures of computer effectiveness are non-existent" (Churchill et al., p. 144).

Data from the Bay Area study discussed earlier confirm the Churchill interview findings through the use of different research methods. Table 1.4 shows the distribution of reports as coded by the research team. As mentioned earlier, during coding it was necessary to add the categories of "none" and "execution of a rule" since too many of the decisions involved could not even be classified as operational control in nature. Table 1.4 shows that most of the reports in the Bay Area systems were historical or were developed for systems maintenance. (Almost all commercial information systems have various editing and control components; these reports are referred to as "systems maintenance.") It can also be seen that, for the information systems in the sample, the decisionmaker was generally a clerk, foreman, or an accountant; only 11% of the reports went to middle or top management personnel. There were some summary and exception reports, though most of the exception reports were for systems maintenance.

TABLE 1.4. VARIABLES AND PERCENT OF REPORTS
IN EACH CATEGORY

Variable	Categories	Results
		% Reports in Each Category
		0 10 20 30 40 50 60 70 80 90 100
Decision type	None	11
	Execution of rule	64
	Operational control	14
	Managerial control/strategic planning	10
Purpose of Report (complexity of problem)	Product	8
	Historical	31
	Maintenance of system	23
	Problem/action	12
	Cost	22
	Planning/trend	5
Decisionmaker	None/outside company	4
	Clerk	27
	"Super" clerk	19
	Foreman/supervisor	16
	Accountant	24
	Middle/top management	11
Report frequency	Daily	16
	Weekly/semiweekly	33
	Monthly/semimonthly	33
	Other (quarterly, yearly, on request)	18
Aggregation (format)	Full report	64
	Exception report	16
	Summary report	20

Both the Churchill and Bay Area studies imply that, in many organizations, computer systems have been used primarily for clerical work and for some operational control applications. There clearly has been little or no impact on the management in these companies. This highly significant result was reported by both studies, though each used radically different research techniques.

CREATIVE DESIGN

As more mundane applications are implemented, there will be a movement toward the development of more decision-oriented systems and new applications affecting multiple functions and departments at one time. This trend can be forecast from the Gorry–Scott Morton framework and is supported by the Churchill study. However, user surveys have shown widespread dissatisfaction with many existing information systems. If we have not succeeded in designing structured operational control and transactions processing systems, how can we hope to design more ambitious and sophisticated systems successfully?

Clearly new approaches are needed to improve the design of all systems and to meet the challenges of designing more sophisticated applications than currently exist in most organizations. The philosophy of creative systems design was developed to meet these challenges. Creative design techniques are focused on solutions to organizational behavior problems in systems design. An essential component of creative design is to consider design a change process and to plan for it. Planning should include considerations of the change process, itself, along with the possible effects of a system on the organization, work groups, and individuals. Creative systems design emphasizes the role of the user in designing systems; we shall suggest that the user should actually design the system himself.

Systems designers have to help the user translate his require-

ments into a workable information system. The interface between the user and the system must be carefully considered. This
physical interface with a system is very important in gaining
user acceptance of a system. As a part of its general user orientation, the philosophy of creative systems design also stresses
evaluating the quality of a system according to users' criteria. If
systems design is treated as a planned change, if the interface
between user and system is considered in design, and if the

FIGURE 1.2.   OVERVIEW OF TOPICS. SOLID LINES INDICATE MAJOR
RELATIONSHIPS; BROKEN LINES, MAJOR INTERACTIONS. (NUMBERS
IN PARENTHESES INDICATE THE CHAPTER IN WHICH THE TOPIC IS
DISCUSSED OR INTRODUCED.)

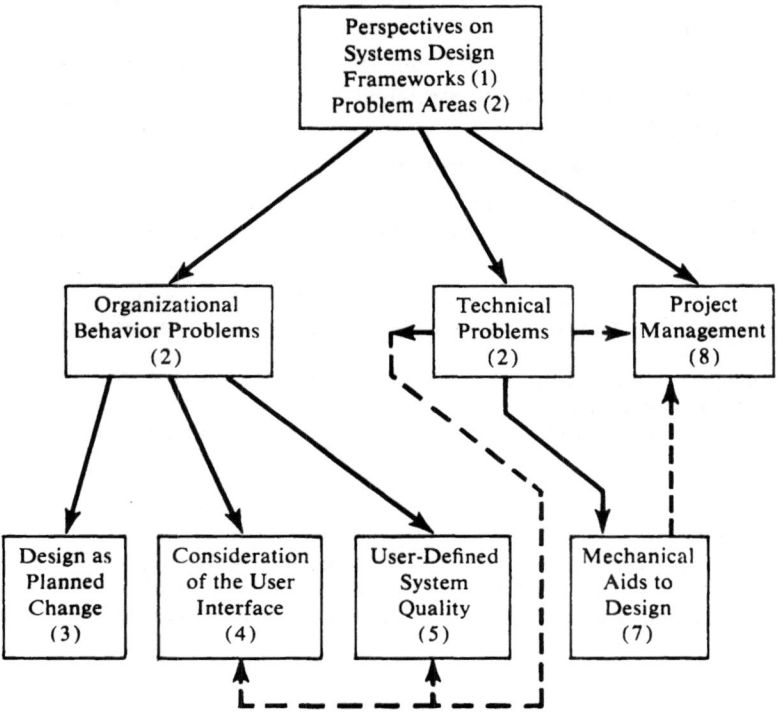

quality of the system is defined according to user criteria, organizational behavior problems in design should be minimized.

OVERVIEW OF TOPICS

Figure 1.2 presents an overview of topics to be discussed and
the relationships among them. In this chapter we have discussed
the motivation for creative systems design, developed a definition of information systems, and discussed frameworks for communicating about systems and predicting the future direction of
systems development efforts. In the next chapter we discuss
three major problem areas in systems design. The philosophy of
creative systems design is focused on solutions to organizational
behavior problems; Chapters 3 through 5 deal with systems design as planned change, considerations of the user interface with
systems, and user-defined measures of system quality. Chapter 6
presents three examples of how creative systems design techniques have been applied in actual situations.

Chapters 7 and 8 deal with technical problems which have
a direct impact on creative systems design efforts. In Chapter 7
various mechanical aids to systems design are discussed; these
aids can help automate some of the technical problems of design
and thus free time and manpower to pursue creative design
activities. Chapter 8 discusses various issues in project management. Without the successful management of the behavioral
and technical aspects of a systems design project, a creative
approach to design will not be effective.

RECOMMENDED READINGS

Anthony, R. 1965. *Planning and Control Systems: A Framework for Analysis.* Boston: Division of Research, Graduate School of Business Administration, Harvard University. (An easy to read book which contains many
   insights on managerial decisionmaking.)

Churchill, N.C., J. H. Kempster, and M. Uretsky. 1969. *Computer-Based Information Systems for Management: A Survey.* New York: National Association of Accountants. (A thorough, empirical survey of computer-based information systems showing their relatively small impact on management.)

Gorry, G. A. and M. S. Scott Morton. 1971. "A Framework for Management Information Systems," *Sloan Management Review,* 13 (1):55–70. (This paper presents a framework which synthesizes the work of Anthony and Simon; the result is a very appealing conceptual view of information systems.)

Lucas, H. C., Jr., K. W. Clowes, and R. B. Kaplan. 1973. "Frameworks for Information Systems," Stanford: Graduate School of Business Research Paper No. 142. (A survey and an evaluation of six information systems frameworks and an example of the use of two frameworks in a systems review.)

Simon, H. 1965. *The Shape of Automation for Men and Management.* New York: Harper and Row. (This book presents Simon's basic approach to the study of organizations and decisionmaking; it is clearly written and is an important contribution to the study of organizations.)

CHAPTER TWO

MAJOR PROBLEMS IN SYSTEMS DESIGN

INTRODUCTION

MOST BOOKS AND RESEARCH on systems design have focused on the technical aspects of design, though some recent works have emphasized the management of design projects. There really are three major problem areas in systems design: technical, organizational behavior, and managerial. Almost all work to date has ignored the middle category except for exhortations on "user involvement" or "participation." The issues here are far more significant than mere user participation. This chapter examines each of the three problems separately, though there is interaction among them.

TECHNICAL PROBLEMS

The most frequent topics addressed in the computer and systems design literature are technical in nature. Hardware and software constraints limit what can be accomplished in any information system. However, given the computer systems available today, a wide range of design alternatives exists. For systems designers who plan to use available equipment and software, technical problems usually begin after the basic logic of the system has been developed.

The design of appropriate record layouts and data structures is a major technical problem for the systems development staff. Once the data storage requirements have been specified, they must be mapped onto physical file devices. Similarly, input and output requirements must be specified, and it may be necessary to assess present machine capabilities to determine if additional equipment is needed. Depending on the nature of the system, it may also be necessary to develop systems programming specifications. For example, the development of an on-line system may require systems programming for the telecommunications interface. The software requirements for applications and any systems programs must be developed and the interface among programs must be specified. Detailed documentation should be prepared and program and system testing planned. Massive amounts of data may have to be collected and special programs written for conversion into a new system.

Our interest in technical issues will be limited to those issues which interact with organizational and project management problems in systems design. We shall ignore technical problems which have no bearing on these other two areas, because technical issues have been addressed extensively in the computer science and engineering literature. Technical problems in determining the logic of the system, input and output design, testing, and documentation will be discussed as they relate to organizational behavior problems. The issues of software design, documentation, and testing will also be discussed under mechanical aids to design and project management.

ORGANIZATIONAL BEHAVIOR PROBLEMS

Our contention is that organizational behavior problems in systems design are the most important. Organizational behavior problems emanate from the impact of the design process and final system on the organization, the work group, and the individual. An information system necessarily creates change. Unfortunately, we have failed to recognize the impact of change on man or dismissed it as "people problems," which are not in the domain of systems design. Yet it is people who are confronted with change and people who cooperate to use an information system.

CHANGE AT THE ORGANIZATION LEVEL

Information systems have the potential to alter radically the power reationship among departments in an organization. Early studies of the organizational impact of computers did not include considerations of power, but instead focused on structural variables. For example, one interest in these early studies was whether or not computer systems resulted in more or less organizational centralization. Unfortunately, in these early studies centralization and decentralization were not well defined (Whisler, 1970). Centralization and decentralization questions are not particularly relevant in trying to understand potential problems at the organization level arising from the design and implementation of information systems. Other studies examined changes in organization structure at the department level. However, there is no reason why one would expect a major consistent impact from computer systems on departmental structure. Changes in departmental structure will be highly dependent on the nature of the computer application and the original structure of the department. (See Lucas, 1973a, Chapter 13, for a summary of some of these studies.)

The fallacy of early studies and predictions of the impact of systems on the organization is that the most important variable has not been included: researchers have looked primarily at structural variables, such as reporting levels, which are easy to measure. The real problem resulting from the design of information systems at the organizational level is associated with an intangible variable, power. From an organizational standpoint, power may be defined as "the determination of the behavior of one social unit by another (Hickson et al., 1971, p. 218)." Different subunits in an organization influence each other; some are more powerful than others. Power affects the relationship among departments and the way in which they carry out their assigned duties. Changes in power relationships alter the way in which departments cooperate; the design and implementation of information systems result in a transfer of power from user departments to the information services department. If the user department resists this transfer, information systems may become victims of the user's struggle to retain power. A descriptive model proposed by Hickson et al. (1971) helps in understanding this problem in the design of information systems. The model hypothesizes that the amount of a subunit's power depends on four independent variables.

According to the model, the more a subunit copes with uncertainty for other subunits, the greater is its power. Uncertainty is defined as the lack of information about future events which tends to make alternatives and outcomes less predictable. There are many uncertainties in the systems design process; it is not clear that the user is aware of his needs or that a system will meet them. Uncertainties also abound in the operation of existing systems; the user is dependent on the information services department to operate an application. The information services department, on the other hand, depends on users to supply accurate input on schedule. The information services department is also in the business of supplying information to users.

One of the major functions of information is to reduce uncertainty. Thus the information services subunit, by its very nature, tends to cope with uncertainty for other departments, particularly when it provides outputs like forecasts and sales histories.

When a computer system is being developed to replace manual methods or to supplement the information provided to a decisionmaker, new uncertainties are created. Before the development of the computer system, the user had almost complete control over the situation, even though his information may have been limited. As a computer system is designed, the user becomes dependent on the information services department for processing. If the user does not understand the system, there is more uncertainty for him after it is implemented. To the extent that the user controlled the process before the development of the computer system, the systems design effort has created uncertainty with which only the information services department can cope.

The power model also predicts that, the lower the substitutability of a subunit, the greater is its power. What alternatives are available to the continued functioning of the subunit, and are these alternatives really viable? For the information services department there are not many alternatives. A service bureau can be used to replace or supplement the information services department. However, management is often unwilling to trust an external organization with proprietary data and the control of its computer systems. Often equipment is not used to capacity and the marginal cost of adding an application within the organization is far lower than the charges for using a service bureau. Another alternative to the existing information services department is to replace personnel. However, programmers and designers make this process difficult through poor documentation. It would also be very difficult to replace the entire systems design and computer operation staff within an organization. A final alternative is to use a facilities management contractor. This consultant contracts to operate the information services department for an agreed-upon price. Frequently he hires many of the existing employees, but is free to substitute his own staff where needed. While facilities management has been successful in some situations, it has the potential for creating many new problems because a separate organization is now responsible for information processing. Given all of these alternatives, it seems difficult, if not impossible, to substitute completely for the information services department.

The power model also hypothesizes that high work flow pervasiveness and immediacy lead to greater accumulation of power by a subunit. Work flow pervasiveness refers to the degree of connection of work flows of one subunit to the next. For the information services department this degree of connection varies according to systems. A particular information system may be highly central and pervasive for a particularly crucial application. For example, the failure of an on-line reservation system can be disastrous for an airline. Work flow immediacy refers to the speed and severity with which output affects the organization. For the information services department, immediacy again depends on the application. It can be very high for a bank which has to process checks and know its reserve position by a certain time each day.

The final variable in the power model is the control of strategic contingencies of one subunit by another subunit. The model predicts that, the greater the degree of control over other subunit's contingencies, the greater is the power. Control of strategic contingencies can also be described as interdependence among subunits. If unit A controls a large number of contingencies for unit B, then B is dependent on A. One of the most unique aspects of the information services department is its heavy contact with a variety of subunits which are dependent on it for service. The quality of these services can be critical for other subunits, depending on the nature of the work flows and the amount of uncertainty involved. Often, when a new system is implemented, functions are reallocated among departments and these changes can create new dependencies among the functional departments and the information services department.

The power model applied to the relationship between the information services department and the other subunits in the organization predicts that power will naturally shift to the information services department as more applications are designed and implemented. This power transfer is an organizational behavior problem which is usually not recognized by the information services department or the user. Each computer application tends to be viewed individually; the cumulative impact of all the applications remains unnoticed.

According to the trends described for information systems and the predictions of the power model, the amount of power held by the information services subunit can be expected to increase over time. In addition to the natural trends in information systems, there is undoubtedly a feedback loop within the power model; powerful areas tend to become more powerful over time. The transfer of power from users to the information services department results in greater dependence on the information services staff. If users resent this transfer of power or try to regain some of the control they have lost, they may not cooperate with the information services department and may not use systems.

CHANGE AT THE WORK GROUP AND INDIVIDUAL LEVELS

A work group has its own norms and rituals and can be thought of as a suborganization within the larger organization. For an information system to be successful, the cooperation and support of work groups must be enlisted. If norms arise in the work group which stress not using a system, then the system has little chance of being effective. Information systems are likely to have a major impact on work groups, particularly those involved in preparing input or receiving output from a newly installed system. It may be necessary to change the content of various jobs and group assignments as a part of the design. Information systems designers have often failed to appreciate the importance of the work group to its individual members. It is imperative to consider the social relationship among co-workers when designing a system.

Information systems also affect individuals in addition to work groups. The same potential problem exists here; individuals must be motivated to cooperate in the use of a system if it is to be successful from the standpoint of the organization. There were several early concerns with the impact of computer systems on individuals, especially with the impact of computer automation on the work force. Since most early computer systems automated paper processing, it was feared that systems would cause vast skill changes for clerical personnel and that heavy unemployment might result. A few studies of the impact of computers on individuals have been performed; see Lucas (1973a) and Whisler (1970) for summaries. The results suggest that there have been some instances of a lowering in the skill requirements for jobs, but generally computer systems have resulted in job enlargement. However, it is likely that a computer system will result in increased rigidity, more deadlines, and greater time pressure, especially batch computer. systems. An important conclusion in several other studies is that different workers respond to computer systems in different fashions based on their personal characteristics and situation (Mumford and Ward, 1968). In designing and implementing a system, these personal characteristics such as age, sex, and length of employment should be considered. For example, one study found that younger female bank employees were relatively unconcerned about the introduction of an early computer system in the bank. The greatest resistance to change in this study came from older male employees who had been with the bank for a number of years (Mumford and Banks, 1967).

One undesirable outcome of group or individual dissatisfaction and hostility toward information systems is conflict between individuals. This problem is particularly severe if the conflict occurs between users and the information systems staff; conflict can lead to a lack of cooperation in designing and operating systems and even to sabotage. A descriptive model of organizational conflict has been proposed which enumerates the different conditions leading to conflict (Walton and Dutton, 1969). Understanding these conditions can help to reduce conflict within the organization. While the relationship between users and the information services staff has a potential for meeting all of these conditions for conflict, it is unlikely that in any one organization all of the conditions would actually be met.

The first condition for conflict is mutual dependence, which forms the link to the power model since dependence is almost identical with the control of contingencies. The process of developing information systems makes users interdependent with individuals in various subunits. The information services department depends on users to help design a system and to supply data for the system once it becomes operational. Users are clearly dependent on the information services staff to understand and translate their needs into processing algorithms. Users also depend on the information services department for the operation of systems once they have been implemented.

Another condition leading to conflict is asymmetrical work relationships; the information services staff has to understand users' jobs, whereas the reverse is not always true. Often misunderstandings arise between users and the information services staff because of the nature of computer work. It is not unusual for computer professionals to keep very odd hours. The user may see the programming staff arriving at 2:00 in the afternoon and not recognize that they have been working all night. Another problem arises because of the task force nature of design which is atypical of most modern organizations; the user is accustomed to a bureaucratic, hierarchical departmental structure. The camaraderie that develops among design teams and the more creative and flexible atmosphere demanded by the task may create conflict and misunderstandings. There are certainly different performance criteria and rewards among users and the information staff that can lead to conflict, according to the model. In general, the basis for organizational rewards is not clear. However, it is safe to say that the information services department is usually not rewarded for working with users. Users also are not rewarded for their cooperation with the development of a computer system.

Differentiation or specialization can also lead to conflict. Systems design and programming are highly differentiated tasks which, as described above, often result in an unusual organizational structure for the design team. The specialization in turn leads to further dependence by the user. The conflict model also predicts that role dissatisfaction can create problems in an organization. It is very difficult to manage computer professionals, and a user is not always clear about the role of the information services staff. Thus there may be dissatisfaction with the computer professionals' role on his own part and on the part of users. Users may not be happy with their relationship with the information services staff, and this is another source of role dissatisfaction.

Ambiguities can also lead to conflict, according to the model. Ambiguities arise naturally as a result of the uncertainties already described as being a part of the design process. The responsibilities for undertaking different parts of the design and the understanding between the users and the information services staff all present many opportunities for ambiguities to arise. Dependence on common resources has also been known to create conflict in organizations. Users may see the information services department as taking resources which would otherwise belong to the user department. Depending on the general organizational climate, this competition can create conflict.

Obstacles in communications have been a major cause of

conflict in the organization; historically this has been a significant problem with the information services staff. The computer has been responsible for an entirely new jargon and, unfortunately, computer professionals may become accustomed to communicating with each other in terms which are unfamiliar to users. In conversations with users, these technical terms may be used by the computer staff member who forgets the users' inability to understand them. Unfortunately, jargon is sometimes used to confuse users or make excuses for a problem. The computer staff may also be confused by special jargon in the users' functional area. Closely associated with communications obstacles is another conflict condition, differences in personal skills and traits. The information services staff has a different orientation and education from those of users. This difference combined with unusual work habits and schedules is a source of conflict in the organization.

As pointed out in the beginning of this discussion, it is unlikely that one organization encounters all of these conditions for conflict. It should also be noted that not all conflict is dysfunctional; in some cases constructive conflict can be healthy, as we shall see in a later example. However, destructive conflict is of concern; it is especially serious because of the crucial interdependencies among users and the information services department. It is relatively easy for a user to sabotage a computer system by supplying incorrect data, omitting transactions for processing, or consistently failing to meet schedules. The information services department can sabotage the user through intentional errors or delays in processing and many other subtle techniques. When this type of conflict becomes rampant in the organization, there is little hope that the various warring groups will be able to eliminate it. Instead, one can expect the information services department and users to spend most of their time assigning blame instead of correcting problems. Under these conditions, systems will not be used if they are implemented at all.

IMPLICATIONS

Designing an information system is an activity which is critical to the long-range success of the organization. It is possible to treat the operation of existing systems as a production problem. However, it may not be possible to operate a poorly designed system at all; users may refuse to cooperate in its use, or the system may be inherently incapable of performing its required functions.

As described above, the user department transfers power to the information services department during the systems design process. This transfer of power is especially noticeable after the system is operational and the user becomes dependent on the information services department for processing. The user department becomes increasingly more interdependent with the information services subunit; work flows are modified and job content is changed for numerous individuals. During this design process there are abundant opportunities for conflict to develop; there is high task dependence in obtaining data from users and in designing the system. There are many ambiguities and communications obstacles as each group learns about the other. Ways must be found to make the transfer of power palatable and to reduce conflict before it becomes destructive. The relationship between users and the information services department actually begins during systems design.

However, systems design does more than inaugurate the relationship between users and the computer staff; it is also a part of the change process of implementing a system and it should be recognized as such. If the change aspects of systems design are not considered, users may become alienated before the system is implemented and may refuse to use it after installation. If no one uses the system then it has failed, whatever its degree of technical elegance and sophistication. Repeatedly, on user surveys, we find dissatisfaction among users and evidence of reports being ignored or thrown out without being read. Though these systems may work technically, they have failed to provide tangible benefits for the organization and have certainly not achieved their potential. The philosophy of creative systems design is devoted to some solutions to the organizational behavior problems presented in this section. However, before these solutions can be discussed, it is necessary to introduce the third major problem area in systems design, project management.

PROJECT MANAGEMENT

Project management encompasses the technical and the organizational problems of systems design. Unfortunately, there has been little research into the management of systems development projects and most work deals with the technical aspects of design. The research and development aspects of systems design make it an extremely challenging management process; in many instances, systems have failed because of a lack of management. Aaron described a number of major systems failures within the IBM Federal Systems Division. His analysis indicates that most of these problems are managerial in nature (Buxton and Randell, 1969).

PROBLEM AREAS

A major problem in systems design has been the failure to specify overall objectives before undertaking a system. What are the goals of the system? What functions is it to perform? Other problem areas have been pointed out by several authors (Jones and McLean, 1970). There is no definition for a good programming code. Should programs be efficiently and "tightly coded," or should they be readable and quickly debugged? Since general standards have not been specified, it is impossible to evaluate the output of a programmer. This lack of knowledge about programmers and the programming task is one crucial management question to be addressed. How should we organize and manage programmers (Kay, 1969; Weinberg, 1972a)?

Another topic in project management is the nature of a project plan and the milestones which should be included. There are also very few standards for developing specifications. Standards are lacking for documentation, and programmers and analysts do not like to prepare it. Documentation, however, is invaluable from a management standpoint for modifying a system and correcting undiscovered errors after the system is operational. Specifying the type of testing to be employed is also a difficult problem; there are a combinatorial number of paths in any program and complete enumeration is impossible. Finally, how does one measure the system against the real world in order to determine its external validity? These uncertainties combine to make the development of a plan a difficult task.

A number of very serious problems in managing the development of information systems have been described above. Unfortunately, not enough is known as yet to indicate which are the key problems. In a seminar described by Kay (1969) a number of project managers discussed why their particular systems had run into difficulty. The most interesting aspect of this conference was the wide variety of reasons provided for failure. One could conclude from this seminar that coping inadequately with any of these problems can result in an unsuccessful system. In Chapter 8 we return to the topic of project management and attempt to offer solutions to some of these problems.

SUMMARY

In this chapter we have delineated three broad problem areas associated with systems design and have described the major issues within each category. See Table 2.1. These three categories

TABLE 2.1.   MAJOR PROBLEM AREAS IN SYSTEMS DESIGN

Technical problems
   Data and file structures
   I/O requirements
   Assessing machine capabilities
   Program specification
   Debugging and testing
   Documentation
   Conversion and implementation

Organizational behavior problems
   Change at the organization level—power transfers
   Change at the work group and individual levels—conflict

Project management problems
   Defining objectives
   Evaluating program and system quality
   Personnel management
   Testing
   Project plans
   Project monitoring

include technical, organizational, and management problems. The three categories are not distinct, and there is much interaction among them. For example, technical problems are related to the organizational problem of designing high quality systems and to a number of the issues in project management. However, the categories help to organize the major problems in design and provide a conceptual overview for subsequent chapters.

The early emphasis of research has been on technical problems in systems design; some recent work has focused on managerial difficulties. It is necessary, however, to consider all three problem areas. Elegant solution in one area and poor performance in another do not suffice; each category of problems is critical. The remainder of this book deals with suggested solutions to some of these problems and presents examples of how some of these solutions have been successfully employed. The remaining chapters are dedicated to the idea that systems can be designed which: are technically sound; fit within the organization; and provide useful results for the user and the organization.

RECOMMENDED READINGS

Hickson, P. M., C. R. Hennings, C. A. Lee, R. E. Schneck, and J. M. Pennings. 1971. "A Strategic Contingencies Theory of Organizational Power," *Administrative Science Quarterly,* 16 (2):216–29. (A theoretical paper presenting a general model of power at the departmental rather than individual level.)

Lucas, H. C., Jr. 1973. "The Problems and Politics of Change; Power, Conflict and the Information Services Subunit," in F. Gruenberger (ed.), *Effective vs. Efficient Computing.* Englewood Cliffs, New Jersey: Prentice-Hall. (An application of power and conflict models to the relationship between the information services subunit and other departments in the organization; system design as a planned change is discussed in detail.)

Walton, R. E. and J. M. Dutton. 1969. "The Management of Interdepartmental Conflict: A Model and Review," *Administrative Science Quarterly,* 14 (1):73–84. (A theoretical perspective is presented on the conditions which have the potential for creating conflict among individuals in an organization.)

Whisler, T. L. 1970. *Information Technology and Organizational Change.* Belmont, California: Wadsworth Publishing Co. (A short and readable review of the impact of information systems on organizations and some of the resulting implications.)

PART TWO

SOME SOLUTIONS TO ORGANIZATIONAL BEHAVIOR PROBLEMS IN SYSTEMS DESIGN

CHAPTER THREE

DESIGN AS PLANNED CHANGE

INTRODUCTION

THIS CHAPTER presents the first component of our creative systems design philosophy, treating design as planned change. In the preceding chapter we discussed organizational behavior problems which emanate from the change process of systems design. The actual design and the final information system modify conditions for the organization, work groups, and individuals. At the organizational level a change strategy needs to reduce the transfer of power from user departments to the information services department. For the work group and individuals a strategy has to take into account group norms and individual tasks to avoid creating conflict between users and the information services staff.

The emphasis of our strategy is on planning an approach to systems design before the actual design effort begins. Instead of looking at conversion and implementation of the information system as the time when changes occur, our philosophy considers the entire process of design as a change effort. The strategy we shall advocate is based on user participation in systems design to the point that the user is actually designing the system. The information services department staff acts as a catalyst helping to translate user requirements into the technical specifications for an information system. Popular data processing literature has encouraged limited participation but has not emphasized it sufficiently, indicated why participation is important, or suggested how to achieve it. We shall try to offer evidence why the user should design his own system and suggest how this approach can be implemented. However, before we present our strategy, it is useful to examine traditional approaches to change to see why a new approach is needed.

TRADITIONAL APPROACHES TO CHANGE

Whisler has described the major change strategy used in the past for implementing information systems as "rational bureaucratic" (Whisler, 1970). The assumptions underlying this change strategy conform with Schein's model of "rational-economic" man (Schein, 1965). This model suggests that man is motivated by economic incentives alone: these incentives are controlled by the organization, and man is a passive agent to be manipulated and controlled by the organization. The model also stresses that man's feelings are irrational; they must be prevented from interfering with rational self-interest.

On the basis of this primitive model, the strategy is to use formal authority to implement information system changes. Changes in organizational structure, work groups, and individual job content are dictated by the organization. The only training provided is on the technical use of the information system. This strategy assumes that organization members will cooperate because it is in their best interest to do so and because they can easily be replaced.

This rational change strategy ignores power transfers among

organizational subunits and the effects these transfers may have on individuals. Changes in work flows and departmental relationships are dictated with no consideration of the consequences. The rational bureaucratic change strategy also heightens conflict. Dependencies among users and the information services department are not considered. Because changes are made arbitrarily, users are not helped to understand the reasons why changes might be desirable. Users are also likely to feel that the information services department does not understand their problems. Without mutual understanding, ambiguities and communications obstacles will increase. The most likely result of using the rational bureaucratic change strategy is resistance to a new system which, if severe enough, can render the system completely ineffective.

Schein describes several more enlightened models of man: Theory X, Social Man, and Self Actualizing Man (Schein, 1965). On the basis of the evidence for each of these models, Schein suggests a model of man called "complex man" which integrates all of these more recent theories. Complex man is highly variable; he has a psychological contract with the organization based on his needs and organizational experiences. The model recognizes that man can have many motives for becoming a productive member of the organization. The model predicts that man will respond in different ways to various managerial strategies according to his unique characteristics. Clearly, a change strategy other than rational bureaucratic is needed if we accept this more appealing model of man.

NEED FOR CHANGE STRATEGIES

A number of different change strategies have been reviewed by Katz and Kahn (1966). In studying these strategies one can conclude that it is necessary to consider the technical nature of the organization (work assignments and other organizational variables) along with the social structure of the organization. In accordance with the theory of complex man, there are a variety of variables to which organizational members respond, including those associated with the organization, personal factors, social relations, and peer groups. For the changes created by the development of an information system, an examination of change strategies suggests that the technical approach of the rational bureaucratic strategy cannot succeed. Instead, we need a change strategy which considers the impact of the design process and the final system on the organization, work groups, and individuals.

Three studies of participation in noninformation systems projects are described below. These studies illustrate the importance of participation in making changes and indicate the crucial role that work groups play in influencing cooperation. While there is always danger in generalizing from one organizational context to another, these studies present compelling evidence of why a change strategy including participation is important for developing information systems. A change strategy and techniques for securing participation in the information systems development process are then explored.

COAL-MINING STUDIES

Schein discusses some of the work of Trist and the Tavistock Institute in England (Schein, 1965; Trist, 1963). These studies deal with the impact of a technological change on a group of coal miners and indicate the importance of considering group norms in changing the work environment. The study was stimulated by the installation of mechanical coal-cutting equipment and conveyors in a process known as "the longwall method." The old system of mining involved small groups of 2 to 8 workers performing as an independent team. Often these groups were isolated from other workers in the mine and long-term relationships tended to develop among the teams; for example, teams cared for a member's family when he was hurt. Strong emotional ties developed among the men in this hazardous work.

The longwall method required radical changes in group structure; small factory departments of 40 to 50 men under a single supervisor were established. The men worked three shifts and each shift had to be coordinated with the others; performance in one shift was dependent on the performance of the preceding shift. In this new system the men were separated more physically than under old methods. Many difficulties resulted from the change, including the disruption of the original groups, and a norm of low productivity arose as a result.

In this instance rational engineering demands had disrupted the social organization so that the new system did not work effectively. To solve these problems a new form of physical work organization was arranged so that the men could form groups to meet their emotional needs. This experience led Trist and his associates to describe change as a "sociotechnical" process as opposed to the original engineering model of change. The researchers stress the importance of looking at both technical and social variables in making work-related changes.

THE HARWOOD STUDIES

Tannenbaum (1966) has summarized the results of a classic experiment in the Harwood Pajama Manufacturing Company. (See also the paper by Coch and French in Proshansky and Seidenberg, 1965). This study provides evidence that group norms can be developed through participation that contribute to rather than inhibit a change effort. The pajama factory had an almost completely female labor force; in general, industrial relations were good except for the chronic need to alter products and production processes in response to market pressures.

Coch and French conducted an experiment on the job change process involving four groups of workers. The control group made the necessary changes in the regular way: jobs were timed and new piece rates were set by the engineers. In the second group, participation was elicited through representatives.

The representatives were convinced of the need for a change and learned the new jobs. Piece rates based on performance were established, and then the representatives trained other workers. In the third and fourth groups all members designed the new jobs and piece rates together.

The results in terms of productivity differed drastically among the groups. The control group dropped from about 60 units of production before the change to 50 afterward and remained there. Productivity for the group with special representatives dropped immediately after the change but after 13 days reached a production level of 60 and after 32 days reached almost 70. The two groups with total participation achieved productivity of over 60 units after 4 days and over 70 units shortly thereafter. In the more participative groups the researchers found less aggression toward management, decreased turnover, and greater identification with work and the company.

AN ABSENTEEISM STUDY

Lawler has reported two studies which support the conclusions of the Harwood experiments; through participation, group norms can be developed which are consistent with the goals of the change agent. The most interesting aspect of the studies described below is that the norm developed through participation persisted a year later despite major changes in the composition of the work group. The studies were aimed at reducing absenteeism among part-time workers cleaning buildings in the evenings (Lawler and Hackman, 1969; Scheflen et al., 1971).

Three separate work groups developed their own pay incentive plans to reward good attendance. These plans were imposed by management on several other work groups; one control group received no treatment and the other participated in a discussion of the problem of attendance with an experimenter. Only the full participation group showed a significant increase in attendance after the implementation of the plans. The authors suggest that possibly participation led to more commitment, that those developing the plan had more knowledge about it so the plan was more suited to their needs, and/or participation increased employee trust of the employer's intentions.

One year later a follow-up study was conducted (Scheflen et al., 1971). During the year the company had dropped the incentive plan in two of the three participative groups; attendance stayed high in the group continuing the plan and dropped below the pretreatment levels in the other groups. A crucial finding was that 60% of the original workers in the participation group had left the company by the time the second study was conducted. However, a norm had been established which was adhered to by the workers. The norm persisted despite the fact that the majority of the group now consisted of workers who had not participated in the development of the plan.

A CHANGE STRATEGY

In this section we attempt to integrate the studies discussed above and to present a strategy for dealing with the organizational changes inherent in the design and implementation of information systems.

PLANNING

As mentioned earlier, a major problem with information systems has been the lack of recognition that the systems design process itself is really a planned change activity. However, it is not enough to recognize the problem; plans must be made for organizational change in advance. Mumford and Ward (1968) have proposed an approach to change which stresses the development of a plan that includes a number of organizational variables. First, it is necessary to consider the environment in which the system is introduced. What is the nature of existing attitudes toward computer systems and the present procedures for accomplishing the organization's objectives? How has the organization handled change in the past? Has it demonstrated a concern for the impact of changes on people? If the climate of the organization is characterized by hostility or conflict, then it is not advisable to attempt to develop an information system. Instead, concentration should be placed on remedial action to improve the existing organizational climate. A total organizational change program is required and outside assistance will probably be necessary. It is hopeless to try to solve organizational problems with an information system. Sufficient problems are created by the information system, and a favorable organizational climate is a necessary precondition to implementing a system.

Second, Mumford and Ward also argue that group relations should be considered in planning. Group loyalties, attitudes, and norms have been established as well as supervisor and subordinate relations. The impact of an information system on these variables should be considered. There also will be inter-departmental relationships which may be affected in some way by the information system. Delineating all of these relationships and establishing a plan for them is an essential part of the systems design process.

Past work has shown that individuals have different reactions to changes brought about by information systems, based upon age, sex, and other personal characteristics. Mumford and Ward advocate developing specific plans, based upon these characteristics, for the different individuals affected. From the earlier discussions, we should also add to the above list, planning for centralization and decentralization. Centralization and decentralization goals should be established by management before the development of the information system.

The focus of this analysis is on the development of a plan which will assist in predicting the short- and long-term consequences of change on the organization, groups, and individuals.

What attitudes are likely to be affected? What behavior can be expected? Where is attention to change needed most? Given this type of analysis, it is possible to plan the change program simultaneously with the development of the system.

THE CHANGE PROGRAM

Given the results of the three studies presented above, a change program specifically tailored for developing information systems should be based on participation. The reasons for involvement are:

1. Participation can be ego-enhancing.

2. Participation can be intrinsically satisfying and challenging (Tannenbaum, 1966).

3. Participation usually results in more commitment to a change (Lawler and Hackman, 1969).

4. Owing to the planning process, participants are more knowledgeable about a change and are better trained; thus user uncertainty about the system is reduced.

5. It is possible to obtain a better solution to the problem because the participants know more about the present operations and functions than the systems design team.

6. Participation means that the user has retained some control over a system which will affect him; he is not transferring as much power to the information services department when he is allowed and encouraged to participate and determine the design of a system.

As with any change strategy, there are problems; Strauss has enumerated several potential difficulties from participation (Tannenbaum, 1966). First, an individual whose ideas are rejected by a participatory group may become alienated. Participation does create group cohesion, but the cohesion may be focused against management and the system. It may not be possible to satisfy the expectations raised for continued participation after the design of the system. Participation is also time consuming and frustrating because progress often seems slow. We should also add to this list the dangers of pseudo-participation, that is, participation which is not genuine. It is better to have no participation at all than to fail to include user suggestions in the design.

Though some of the above-mentioned drawbacks to participation are potentially serious, a well-designed change strategy should prevent them from occurring. Even with these dangers from participation the evidence in its favor is impressive. The studies discussed earlier show that participation can help to establish group norms which contribute to the change process. Systems design inaugurates the relationship between users and the information services staff. If this relationship is antagonistic and conflict develops, the absenteeism study shows that it is likely to persist after the system is implemented. Carefully planned participation should reduce levels of conflict because the user has a say in how the system is developed and specifications are not forced on him. Users also retain more control; thus the transfer of power to the information services department is reduced.

TECHNIQUES FOR PARTICIPATION

Given the advantages of an information system change strategy based on participation, it is necessary to develop specific actions to implement the strategy. The first one is the formation of a high level design team and steering committee. Management commitment to projects must be demonstrated, and this can be best done through management assuming an active leadership role in development of systems. A steering or priorities committee should also be formed with top managers to help make resource allocation decisions. Decisions as to what systems are undertaken should not be left to the information services department. If the information services department must make the decisions on the projects to be undertaken, conflict is likely to develop between the department and users whose requests are denied. Participation in making decisions about systems should help the user to understand the reasons for the rejection of a proposal. A representative group of users is able to choose new applications and enhancements to existing systems based on the overall objectives of the organization.

The second major technique in our change strategy is to decentralize the design process so that the design team includes a large representative group of users. Managers will have to provide their own time and time for their subordinates to participate in design activities. The approach outlined above cannot succeed without this cooperation. Given the steering committees and a decentralized design team, *a goal of the systems design staff should be to let users do the actual design.* This goal can be accomplished through the design team itself, through system review meetings with a larger number of users than are present on the design team, and through several other techniques illustrated in Chapter 6.

One approach that can be used in larger projects where widespread participation on the design team would create an unwieldy group is the "county agent." Whisler (1970) suggests this form of participation because of the time requirements of full participation. However, employing a county agent or user representative should supplement rather than substitute for the design team. The duties of this individual are to understand the system requirements of users and communicate them to the design team. Ideas from the design team are then carried back to users by the county agent for comment. This technique can also be used successfully for areas which are only peripherally affected by a system.

MONITORING THE CHANGE STRATEGY

To follow the approach suggested above and the planning techniques encouraged by Mumford and Ward, some mechanism

TABLE 3.1.   SAMPLE INTERVIEW AND QUESTIONNAIRE ITEMS

Attitude Questions

Directions: Circle the number which best represents your opinion.
For example: The temperature inside today is:

Too cold _____ Too hot
            1   2   3   4   5 (6) 7

The answer indicates that the temperature is hot.

1. My general impression of the data processing staff is that they

are uninterested in the user	_____ 1 2 3 4 5 6 7	are interested in the user
are not too competent technically	_____ 1 2 3 4 5 6 7	are highly competent technically
are not too good in dealing with people	_____ 1 2 3 4 5 6 7	are good in dealing with people
do low quality work	_____ 1 2 3 4 5 6 7	do high quality work

2. How do you think a computer might benefit you? (You may answer more than one.)
   1. Reduce the time I spend processing papers
   2. Reduce errors
   3. Make my job more interesting
   4. Make it easier to find information
   5. Make it easier to use information
   6. I don't know specifically how, but feel it would help

   Comments _____

3. What do you think the major problems with a computer would be? (You may answer more than one.)
   1. It would make things more complicated
   2. It would make more mistakes
   3. It would be harder to use
   4. It would take more of my time
   5. It would lose information
   6. It would make our jobs boring
   7. I don't know specifically what, but feel there would be problems

   Comments _____

4. If we use a computer here, the company will not need me anymore.
   1. Strongly agree 2. Agree 3. Neutral 4. Disagree 5. Strongly disagree

is needed for measuring attitudes and collecting design data. The most common method of data collection for design purposes is the interview. Designers and users often meet individually or in groups to discuss a new system. Another useful data collection method is observation, particularly for more sophisticated decision systems. The observer merely watches and notes the decision process without interacting with the users being observed.

Data collection should also occur periodically during the design process to monitor crucial organizational variables. Such data collection is particularly important in view of the role of the systems designer as a change agent; information is needed to evaluate his change activities. Lucas (1971a) has proposed questionnaire and interview techniques for this data collection purpose. These procedures facilitate the collection of data on present user attitudes and expectations about computer systems. See Table 3.1 for an example of some of these additudinal questions. In addition, data are collected for actual help in designing the sys-

TABLE 3.2.   SAMPLE INTERVIEW AND QUESTIONNAIRE ITEMS

Design Information

Please rate the following reports on the indicated characteristics

Inventory Status Report

1. Highly accurate _____ Highly inaccurate
   1 2 3 4 5 6 7
2.   Out of date _____ Timely
   1 2 3 4 5 6 7
3.   Useful _____ Useless
   1 2 3 4 5 6 7

Sales Analysis Report

4. Highly accurate _____ Highly inaccurate
   1 2 3 4 5 6 7
5.   Out of date _____ Timely
   1 2 3 4 5 6 7
6.   Useful _____ Useless
   1 2 3 4 5 6 7

tem as shown in Table 3.2. It is recommended that these measures be used at intervals throughout the design process to monitor organizational variables. The information collected can help assess whether the change plan is succeeding and whether it needs to be revised. Periodic monitoring provides feedback on the process of systems design and the change effort. After the completion of the implementation phase, a post-installation audit using the questionnaire and interview procedures is also recommended.

SUMMARY

In this chapter we have presented the first component of the philosophy of creative systems design. Systems design is treated as a planned change process from the inception of the system. Planning the impact of the system and changes means that the organizational behavior problems of systems design will be consciously considered. The impact of the design and system on the organization, work groups, and individuals should be a part of

TABLE 3.3. A CHANGE PROGRAM FOR SYSTEMS DESIGN

Planning for
  Impact on the organization
  Impact on work groups
  Impact on individuals

The change strategy
  Participation
    Management committee
    Representative design team
    Users actually design system
    County agents
  Monitoring the plan: collecting design and attitudinal data
    Observation
    Interviews
    Questionnaires

the plan; potential power and conflict problems should be considered. The recommended change strategy is based on user participation to the extent that the user actually designs the system. Participation in the design process helps to create norms which support the change. Also, if the user is in control of systems design, he transfers less power to the information services department and there is less chance for conflict to develop with the information services staff. A method for monitoring the change program and for data collection was also described. The major points in the chapter are summarized in Table 3.3. In the next two chapters we discuss two other essential components of systems design: consideration of the user interface and a user-oriented definition of system quality.

RECOMMENDED READINGS

Katz, D. and R. L. Kahn. 1966. *The Social Psychology of Organizations.* New York: John Wiley and Sons. (The chapter on organizational change summarizes a number of useful approaches to change; further readings are also suggested.)

Mumford, E. and T. B. Ward. 1968. *Computers: Planning for People.* London: B. T. Batsford. (An excellent book suggesting strategies for planning the organizational impact of information systems; the last chapter and appendix are particularly relevant.)

Schein, E. 1965. *Organizational Psychology.* Englewood Cliffs, New Jersey: Prentice-Hall. (A short monograph which summarizes much of the work in organizational behavior and develops Schein's model of "complex man.")

Tannenbaum, A. S. 1966. *Social Psychology of the Work Organization.* Belmont, California: Wadsworth Publishing Co. (Another short book which describes many of the findings of social psychology which are applicable to planning the work environment.)

Whisler, T. L. 1970. *Information Technology and Organizational Change.* Belmont, California: Wadsworth Publishing Co. (In addition to discussing studies of the impact of computers on organizations, this book contains a good section on change strategies.)

CHAPTER FOUR

CONSIDERATION OF
THE USER INTERFACE

INTRODUCTION

DESIGN OF THE INTERFACE between the user and the information systems is a crucial part of creative systems design. The user interface discussed in this chapter does not encompass logical capabilities as much as it does the input/output interface and manual procedures associated with the system. The user interacts with any system whether it is batch or on-line; he is often unaware of files or processing algorithms. To the user, the entire system consists of the input he provides or the output he receives. It is also important to remember that different users have different goals in using a system. The range of possible reasons for using the system needs to be considered in designing the user interface.

A well-designed physical user interface can contribute to the solution of organizational behavior problems in systems design. First, a high quality interface is easier for users to accept and there is a greater incentive to use the system. If the interface is well designed, changes in existing procedures will clearly be an improvement for the user and his work group and the impact of the system should be perceived as favorable. A good user interface means there will be less need for the user to interact with the information services department under adverse circumstances, and thus the potential for conflict will be reduced.

There is evidence that users and the public are becoming much more concerned and less enthusiastic about information systems as a result of problems with the physical interface. An example is given by McCracken (1971) of the difficulties encountered by a computer professional in dealing with a credit card system. The credit card holder was charged with an airline ticket from Denver to Montreal; the charge slip had his account number but another signature on it. The card holder wrote what he described as an "emotionally neutral" letter explaining the mistake, which was not answered for some seven weeks. In the meantime, he received past due notices and threatening letters. He sent more letters and received more threatening letters in return, including a "final notice." Finally a computer-signed letter arrived saying that everything had been corrected, but unfortunately it too was followed by more threats. The only way that the card holder was able to get attention was through a letter describing the entire history of the incident, which appeared in the financial section of the *New York Times*. (The financial reporter received more mail on this particular column than any other he had printed in his twelve years with the newspaper.)

In this example the manual procedures which interfaced the computer system with the user are probably to blame. Either the credit card holder's letter was overlooked, there was too small a staff to respond, or procedures for adequate customer service were lacking. Individuals who finally responded to the card holder's letter either did not modify the computer system files or did not verify that the files had been changed to reflect their correspondence. The computer system, itself, continued to produce inappropriate output since it had no way to take the incorrect data on the file into account. Whatever the specific reasons for these problems, the physical user interface was clearly inadequate in this case.

Stories such as this are not unique. Moving causes difficulties in changing all the records containing an address. In one personal instance a company repeatedly sent its bill to our previous address. Each time we filled out the change of address stub on the bill, but to no avail. Finally, in desperation, we stapled a note to the bill showing the current address. Stapling a note to the computer card forced manual attention and the address was finally changed. Unfortunately, three years later we still received promotional data from the company even though the account was closed by the computer system.

These two examples suggest some of the problems seen by customers, but the problems are just as serious within organizations. There are many stories of users throwing away output without reading it; one of the reasons for this is systems design which ignores the physical user interface. A system which alienates users or customers or is not used at all cannot be considered successful.

BATCH COMPUTER SYSTEMS

Batch systems provide an indeterminant around time, that is, the time from which input is submitted until output is received is not known in advance. All the input is collected and the programs in the system are executed once to process this batched input. Usually the processing involves posting transactions and updating one or more master files and the production of a series of output reports which are returned to users. A number of the suggestions below for improving the physical interface are taken from a paper by Withington (1970).

INPUT

Users provide the input for batch systems in a variety of ways; the most common is by filling out a form which is then keypunched. Keypunching has a very low error rate if verification (a proofing process in which the same information is rekeyed and compared with the original punching) is performed by a separate operator. It is interesting to note that, on user surveys, many of the problems associated with accuracy in a system are attributed to keypunchers. Frequently comments like this are made: "Our keypunching staff should be replaced because they are inaccurate." These statements point out that the user's perception of a system is often focused only on the input or output. If the forms are legible and keypunched and verified, very few errors will be introduced into the processing. The problems referred to by these users are probably weaknesses in the control procedures, processing, or the quality of the data that were actually provided.

One of the problems with keypunching and verification or similar processes such as key-to-tape or key-to-disk recording is that the information must be transcribed before processing. While the error rates for this operation are generally low, the potential is present for making errors and the entire process introduces delays in providing the output. A number of techniques have been used to capture data at its source and avoid intermediate transcription without moving to a fully on-line system. Mark sensing is one such technique in which the user provides input by marking boxes with a pencil on a special form. The forms are read by a mark-sensing device and automatically transcribed onto punched cards or are read directly into the computer system. The process is similar to that used in the machine scoring of standardized high school or college entrance tests. In applications where suitable, that is, where there are a reasonable number of mutually exclusive input fields, mark sensing is an inexpensive and fairly rapid method for source data collection. Unfortunately, if there are a large number of input fields, the forms become unwieldy and difficult to mark.

Optical character recognition techniques provide more sophisticated source input. Optical character recognizers (OCR) come in a variety of styles and prices. Some of these devices can input a number of type fonts and certain carefully hand-printed numbers and letters. Other OCR devices are limited to reading a single type style or even to special bar codes such as those found on some oil company credit cards. While these devices have fairly high reject rates, they read with great speed and avoid any intermediate transcription. Recently, a major computer manufacturer has provided these optical scanners for some of its small and medium-priced computer lines; use of this input medium should thereby be encouraged.

An early technique for reducing the amount of input is the turn around document. Information such as the amount due on a bill is punched into a card which is sent to the user. The card is returned with payment and, if the payment matches the amount billed, no additional information has to be entered. The concept of the turn around document has also been extended to OCR. Instead of sending a card, a form with information printed in one of the special OCR type fonts is prepared and sent to the user, who makes necessary modifications and returns it. More information can be provided on a smaller document this way and users are not confronted with punch cards, which some people find offensive.

The primary problems with all batch input are its rigidity and the lack of immediate feedback. The input documents must be handled in numerous stages, and elaborate control procedures have to be devised to be certain that the input is entered correctly. Exceptions and errors are frequent, and they are reported back to users who must then correct them and resubmit them for the next processing cycle. In the meantime, the system is out of date because the errors were ignored in processing. To avoid these problems, systems designers should attempt to provide to users as much precoded information as possible; this saves the user from laboriously writing unnecessary information. Carefully documented procedures for remedial actions are also important so that the user can figure out what is expected of him. For tracing errors it is helpful if the system can provide meaningful feedback suggesting the source of the error. (Output feedback is discussed further below.)

Frequently, with batch systems, a group of employees is responsible for providing input and correcting any errors. For a large number of members of the organization, often at the clerical level, this activity constitutes a major part of the work day. In general, these individuals are not knowledgeable about computer processing and the importance of their input; however, they are trying to do a good job. It is very discouraging for them to receive lists of exceptions and errors day after day. It is more cost effective to let the system catch some of these errors than for persons to try to provide perfect input. However, because this tradeoff is not explained to the individuals involved, a discouraging list of errors is returned each processing cycle. This procedure can lead to a great deal of frustration, and those providing the input may cease trying to perform well. One technique for avoiding this is to provide a benchmark for employees who have to provide input. Explain that no one designing a system expects all of the input to be perfect and that is why the error routines are provided. Let the users filling out the forms know what a good reject rate is so that they will feel some success if they are able to drop below it. Explaining that perfect error-free input is not worth the user's time can help reduce some of the frustrations with the batch–user interface.

From a design standpoint, it can be very helpful to have users work with different input forms on an experimental basis before they become a part of a system. The user involvement should result in a better system and one which is more pleasant to use. Letting the users design the input forms is one type of involvement and also represents an investment in training.

OUTPUT

A computer system can succeed only if it is used; the quality of output reports determines the value of the system to many users. In designing batch output it is especially important to hold numerous review sessions with users. Copies of pro forma reports should be provided and the user should be encouraged to design his own report if possible.

A number of design principles have been suggested; see Withington (1970). First, it is desirable to provide a clean and easily understood format; it should be very clear to the reader what is on the report. Descriptive labels and titles should be used as much as possible. A page should be provided with an explanation of what the data mean and how they have been derived. What should the user do with the information? From some of the questionnaire surveys we have conducted, users feel at a loss when they receive reports with no explanation. They do not understand how the data have been processed or what they are supposed to do as a result of the information on the report.

Output reports also contain numerous error messages, particularly input error control reports for those charged with providing batch input. The user interface is improved greatly if the feedback provided is complete, understandable, and courteous. Today's computer systems provide sufficient storage that it is not necessary to use obscure codes to describe error messages. A full explanation with suggestions about the source of the error is appreciated by the user.

The results of some of the user questionnaire surveys described in Chapter 1 also show that there is entirely too much output produced by batch computer systems. Frequently this output is necessary for legal or historical records; however, large reports are often produced which are infrequently referenced. Several techniques can be employed to reduce some of the problems with the vast amount of data produced on reports. The first question to investigate is whether or not the report can be produced on an exception basis. Can users provide ranges for data which are acceptable? Are historical trends available so that only a deviation from historical data or projected figures has to be printed?

If the full report must be printed for some reason, can an index be provided to reduce the physical inconvenience of using the report? It might even be possible manually to add indexed tabs with keys to the various pages to facilitate finding information. Again, if the full report must be printed, can an alternative output form, such as microfilm, be used? Some installations have begun publishing summaries of information and printing the full report only when it is specifically requested. A final solution is to put the report on-line for inquiry purposes, even though the files themselves are updated in batch mode. For information which does not need to be updated immediately, this technique provides an attractive solution for reducing the amount of data distributed.

ON-LINE SYSTEMS

On-line generally refers to a system in which the updating of files is done immediately when information is received. However the term can be used to refer to any system where there is interactive entry, output and/or updating. There are systems, for example, in which input is captured by a computer, edited on-line, and then batch processed overnight. Another possibility was described above in which a report prepared in batch mode is placed on-line for inquiry purposes. These two alternatives for input and output are generally much cheaper and are less in-

volved than trying to develop a fully on-line updating system
(Lucas, 1973a).

INPUT AND OUTPUT

Many of the same principles discussed for batch input apply
to on-line systems. Again it is not desirable to use obscure
codes; the system should print out adequate explanations of errors and should be courteous. There is no reason why it cannot
use words such as "please" or "thank you" (Withington, 1970).
It is also helpful to keep the user informed about what is happening. For example, during a delay the system may print out
"searching" or a similar message to show that processing is
continuing.

The nature of the physical device and the format of the
command language (or input language) are extremely important in determining the quality of the user interface. One
design goal should be to minimize the amount of typing required,
as most users are not typists. One method for accomplishing this
is to use menu displays; a list of options with numbers is provided and the user only has to respond with the appropriate
number. It is also possible to use function keys in which a single
key can be programmed to input an entire function. Overlays
for the keyboard are provided to simplify entry.

Any on-line system should also provide some form of a
"help key." By simply depressing this particular key, the user
should be able to obtain information about how to use the system. It may even be possible to recognize what the user is attempting to do by the sequence of requests leading to the request
for assistance. It is particularly helpful in on-line systems to build
in different levels of response in accordance with the needs and
familiarity of the user with the system. For the inexperienced
user full textual information may be provided. For an intermediate user possibly only abbreviated prompts need be given,

and for the experienced user very little feedback at all may be provided.

From the standpoint of the physical terminal, it is desirable to use silent and pleasing terminals wherever the budget allows. Cathode ray tube devices (CRTs) have become far less expensive; these devices offer more flexibility and higher speeds and are completely silent. Light pens or cross hairs can be used to permit direct input, a particularly useful feature if graphics are used. It is also easier and cheaper to display full error messages on the CRTs because of their high speed. The major drawback with these devices is the need for hard copy, though a number of manufacturers are providing hard copy options with their terminals.

SYSTEM RESPONSIVENESS

As mentioned above, it is desirable to let the user know what is happening to his requests. In general, the system should be designed for a reasonable response time and limited peak load degradation. One early time-sharing system, CTSS at MIT, refused new accesses when a certain number of users were logged on in order to provide good service to those currently using the system. Here a decision was made to deny additional service rather than to degrade performance for all users by allowing unlimited access. The priority interrupt systems of modern computers should make it possible to provide a reasonable and consistent level of service. An on-line system should not be developed if adequate resources are not devoted to providing the kind of response users expect.

A user representative is helpful for batch systems; such an individual is crucial for an on-line system. There should be someone available at all times the on-line system is functioning who can be contacted by the user if there is a problem. This individual should be an expert in the use of the system and should be able to explain how to solve the user's problem. It is helpful to design the system so that the user representative has his own terminal to see what problems are bothering the user.

GENERAL DESIGN AND TESTING

In addition to the specific input/output interface, there are over-all design considerations which influence the user's interaction with the system. An important activity for the systems designer is to place himself in the position of a user, especially a user who does not understand computer systems, in attempting to design a system.

COORDINATION

Many complaints concern the duplication of information and input and output. Most applications in the organization have grown on a piecemeal basis; there is little long-range planning or coordination among systems. For example, a bank may have demand deposit accounting which shows balances in checking accounts; the same information is reported on daily and monthly statements and even in a budgeting system.

Even within the same applications, it is possible to have duplication, especially with a batch system. The example given at the beginning of this chapter of the difficulty in changing addresses for a charge card probably arose from the presence of two master files. One master file was developed for billing purposes, and it is updated several times a week; a shorter copy of the file containing addresses is maintained for promotional mailings. This second file is updated only once or twice a year and is out of phase with the master file containing the most recent information. This kind of duplication and lack of co-ordination creates problems for customers. The same conditions also produce frustrations for users within the organization when the same information must be provided in a slightly different form to different systems and duplicate and overlapping information is received on output reports. We need to attempt better coordination among systems since the user sees the entire pattern, not just one application.

MANUAL PROCEDURES

Though not a particularly pleasant or exciting topic, manual procedures are extremely important in all types of systems. Manual procedures are activities surrounding input/output operations; examples are customers queuing at a service counter with a terminal or users filling out input forms. It is possible to design a system that is technically successful but which fails because of manual procedures which cannot be followed. One particular inventory system was designed for updating several times a week; however, there were too many error messages which had to be corrected manually by the next input cycle for this frequency of updating. Instead, the system was updated more infrequently and proved less valuable than originally forecast.

The designer must ask a number of questions in considering the manual procedures implied by or specifically designed for the system. Are the input requirements reasonable? Can the data be collected and transcribed if necessary? Is there an incentive to provide accurate data? In subsequent chapters we discuss examples of ill-conceived systems in which there was actually an incentive against using the system. Again the user needs to be considered; for example, salesmen are notorious for their antipathy toward paper work. Does the system stand any chance of succeeding if salesmen are forced to prepare cumbersome and lengthy input forms? Another system that was implemented depended on extremely accurate time-card reporting for cost accounting. However, this information was also used to

provide recommended staff levels and to reduce the staff automatically if work loads changed. Clearly, there was no incentive
for anyone to prepare the card accurately, as it could result in
his own dismissal!

Manual procedures also apply to the output. What does the
user have to do to make use of it? Does he have to perform
a number of calculations that the computer could perform just
as well? Is the report so large that it cannot be separated and
distributed in time when produced by a batch computer system?
One frequent complaint is that reports are universally late, and
this may be because they cannot physically be produced at the
same time. The impact of a new system on the operating procedures of the computer center should be investigated. It may not
be possible to provide the turn around time expected for reports
during peak processing times at the end of the month or quarter.
For an on-line system, are there enough terminals and is the
system available as promised?

TESTING

No matter how carefully a system is tested, conversion problems can be expected (McCracken, 1971). However, this does
not mean that exhaustive testing should not be undertaken. The
more carefully a system is tested, the fewer difficulties will there
be when it is time for implementation. Testing the interface as
well as the logic of the system is important. For example, it is
desirable to run experiments with the input and the output;
users should be asked to try out the forms and to make suggestions for modification if they have not actually designed the
forms themselves.

The need for testing the interface is especially great for
on-line systems; tests should include as many manual procedures
as possible. In one case an airline developed a new reservation
system in which passengers were assigned to seats by a computer

system. This system replaced the old manual technique of using cardboard cutouts which were handed to the passenger. The system was implemented at Thanksgiving, one of the busiest times of the year, and the operators were not familiar with it even when the Christmas rush arrived. The physical interface included a spotlight focused on the cathode ray tube device which made it very difficult to read. Agents who were not typists had not been trained in the use of the system. The result was long check-in lines and delays in boarding. Later, the airline changed the physical setup of the air terminal to increase the number of check-in lines and to eliminate checking in at the gate itself.

Millions of dollars were invested in the equipment and software for this particular system. A very small amount of money would have been required to develop a simple simulation program (possibly using a small time-sharing system) to estimate the queue sizes and the processing rates. In fact, this simulation could have been tested in a terminal by using actual passengers. Millions of dollars were spent, but evidently nothing was devoted to a simple but vital component which was one of the main customer contact points with the system.

CONVERSION

No program is ever fully debugged. The same is true for the process of fitting all the programs together and converting them into operational status. The best approach, in addition to thorough testing, is to recognize that there will be problems and to prepare employees for them. There are a number of ways to minimize the difficulties of conversion, including thorough training. It is desirable to convert in an off peak season and to convert a single section or a function of the system at one time if possible. For example, if an on-line system is being developed, can one geographic area or one part of the organization be brought up at a time? User representatives should be posted for assistance during conversion, and emergency backup must be provided.

Conversion is the time when user involvement really pays off; the burden of conversion falls on users. The information services staff expects disruptions; it must also prepare the users for them. It must explain the problems of conversion and let the users design the conversion plan and procedures. They know their jobs best and they know what will be required in order to implement the new system. If the users have been involved

TABLE 4.1.   PHYSICAL USER INTERFACE CONSIDERATIONS

Batch systems
  Input
    Forms
    Source data collection
    Rigidity and feedback
    Employee reactions
    User design of forms
  Output
    Format
    Explanations of output
    Courteous and complete messages
    Filtering of output, exception reporting

On-line systems
  Input/Output
    Courteous prompts
    Full messages
    Input simplicity
    "Help" keys and levels of prompts
    Pleasant terminals
  Responsiveness

General
  Coordination of system
  Elimination of duplication
  Manual procedures design
  Testing
  Conversion

---

up until the conversion stage, they will be far more familiar with what has to be done and will be better trained and prepared for this crucial activity.

SUMMARY

Considerations in the design of the physical user interface are summarized in Table 4.1.

RECOMMENDED READINGS

Lucas, H. C., Jr. 1973. *Computer Based Information Systems in Organizations*. Palo Alto, California: Science Research Associates. (For some further discussions on design considerations see the section on Systems Analysis and Design.)

McCracken, D. 1971. ". . . but the ambivalence lingers," *Datamation*, 17 (1):29–30. (A short article discussing some of the frustrations the public has encountered in using computer systems.)

Withington, F. G. 1970. "Cosmetic Programming," *Datamation*, 16 (3): 91–95. (A series of excellent recommendations for desirable design features for the physical user interface.)

CHAPTER FIVE

USER-DEFINED SYSTEM QUALITY

REQUIREMENTS

THE INFORMATION SERVICES STAFF often equates quality with the technical sophistication of a system. How elegant is the design of the system and how complex is the programming? Were especially clever techniques used in developing it? Technical quality may be an advance in the state of the art or the development of particularly elaborate file structures or processing algorithms. Technical quality is important because it is essential that the system work; however, quality as perceived and defined by the user is the dominant evaluation criterion in creative systems design.

There are several reasons to encourage the evaluation of system quality based on user criteria. First, if the user is to participate in design, then he must be convinced that his involvement will be productive. If the quality of the system is defined from the user's perspective, participation becomes meaningful. Second, user-defined quality creates less frustration for the user and tends to reduce the opportunity for conflict between users and the information services staff. The user is not confronted with an arbitrary system which does not really meet his needs. Third, if the user is involved and defines the criteria for evaluating the quality of the system, he becomes more familiar with the system and understands it better. Finally, the more control over the system exercised by the user, the less uncertainty and the less power are transferred from the user to the information services department. Under these conditions, there is a far greater chance that the user will accept and use the system.

GOALS AND PRECONDITIONS

There is a serious question whether an information system can succeed when its goal is to control or punish the work force. Since almost any system requires user cooperation, it is unlikely that such a disciplinary system can succeed even if in some way it can be considered to be in the best interests of the organization. Systems which are "policemen" are in direct conflict with the model of complex man in Chapter 3 that we have used to support our philosophy of creative systems design.

Just as it must be considered whether or not the system is well conceived from the standpoint of the users, it is not advisable to design an information system to solve organizational behavior problems. Sufficient problems of this nature occur during the development of the system; confounding these problems with a system designed to resolve some organizational issue may produce impossible requirements for the system. If there is a major organizational behavior problem which is completely independent of the system under consideration, this problem should be solved before undertaking the system. Specialists in organizational development should be responsible for a solution, not specialists in information technology and systems design!

Another reason for solving organizational problems before implementing a system is that users should not be given a reason to perceive a causal relationship between the implementation of a system and an organizational change. The system is then viewed as the stimulus for the organizational change, even though the opposite was probably the case. A factory floor data collection system whose major goal was to prevent over-reporting of production by factory workers is a good example of a system designed to solve an organizational behavior problem. The work force was paid on a piece-rate basis and there was no incentive for them to use the system. Worker attention focused on this one aspect of the system, and any benefits from it were overlooked. Soon after installation, the system was removed, primarily because of the lack of user cooperation.

USER REQUIREMENTS

The criteria for quality should be defined by the organization and the user. These criteria will, in general, differ from those of the information services staff, but they will still provide many challenges. This approach requires analyzing decisionmaker and user needs and developing a design to meet these needs. Clearly, involvement is important; the user must communicate his needs to the designer, who must communicate the system requirements and capabilities to the user. From the data summarized from open-ended questions on the user surveys described in Chapter 1, there is no broad consensus as to what quality means in a system. Below we attempt to describe general criteria for quality which appear to be important to most users. However, there is some evidence that individuals have different approaches to problem solving, that is, different cognitive styles (Mason and Mitroff, 1973; McKenney, 1972). Designers should consider the decisionmaker's approach and his potential uses of the information. The challenge for the design staff is to determine in detail what quality is for each system and each user. Designers must become skillful at building descriptive models of user needs and decision processes.

MEETING USER NEEDS

OPERATIONS

Certainly no system can meet user needs if it does not work properly. While this is a technical issue to some extent, systems have failed because of improper testing or inadequate user acceptance conditions. Are the processing algorithms accurate? Can timing requirements be met from the receipt of input to the production of outputs? Manual procedures for batch systems must be analyzed, and queuing properties of on-line systems, particularly under peak loading, must be considered. Of fundamental importance to the successful operation of the system is the maintenance of file integrity. Can procedures be designed so that files will be properly maintained? Can the input be collected and posted within a reasonable time?

DESIGN ALTERNATIVES

In addition to determining that the system is capable of performing the processing, the analyst should attempt to describe the minimum set of functions which has to be included in the system. It is not desirable to develop a huge monolithic design. Instead, efforts should concentrate on including only the minimum functions along with options for enhancing these minimum capabilities. Design can be regarded as a step function. For the first design, possibly functions A and B are included; another step will have to be taken to reach design two which includes C, D, and E, etc.

The reason for such an approach is to allow the user and the information services staff to choose the appropriate alternatives for the system. These different alternatives will have varying costs and benefits. The user and the staff should be able to choose a basic set of functions plus options according to their needs. Usually, the only choice for the user is to say "yes" or "no" to a single massive system encompassing all options which the information services staff was able to develop. The simpler the system is, the shorter the development time and the better the chance of its working properly. Also, simple systems usually involve fewer organizational changes; for example, see the United Farm Workers system described in the next chapter. The physical user interface described in the preceding chapter is also an important part of the design and operation of quality systems. This is the user's main point of contact with the system, and it is worth spending time to design a successful interface.

DESIGN CONSIDERATIONS

As a system is being designed, its quality will be enhanced if some of the frameworks described in Chapter 1 are used to provide a general overview for design. These frameworks help to classify the general types of decisions and the information required to support them. If the present information does not seem to match the decision, there may be a problem. Possibly the difficulty is with the decisionmaker's own current method of analysis; he may be limited by the information provided. It may be necessary to modify decision processes before designing a system. In general, the framework should be used to test and explore the design of early versions of a new system.

In considering the quality of the design and operation of a system, a word of caution is in order concerning the amount and type of information presented. Ackoff (1967) has written extensively about the myth that more information necessarily means better decisions. The quality of a system and the information it produces will not in general be dependent on the total amount of data provided. Instead, we need to determine what information can be used best by the decisionmaker and provide only this type. There is some experimental evidence that the quality of output decisions decreases if information overload occurs. Unfortunately, the proper amount and complexity of information for each individual is partially determined by his own integrative complexity, and general rules are difficult to derive (Driver and Streufort, 1969). However, these exploratory findings should serve as a warning against providing data because they exist rather than because they are needed.

How do you determine what information is useful to a decisionmaker and how much to provide in more detail than the general guidelines provided by a framework? It is not really possible to ask this question of the decisionmaker and expect him to answer. Instead, the designer must study the decisionmaker at work to see what information he seeks and how he uses it. The design team must build a descriptive model of what is happening now. Then iterations with the decisionmaker are required to revise the model of his decisionmaking and to see how various changes might effect his actions. The design team should provide samples of new or different forms of information to see how the decisionmaker reacts (Scott Morton, 1971).

After working with an individual decisionmaker, the effort should be expanded to include research on a larger sample. For example, research is currently under way on the use of information by the sales force and how use relates to performance in a national company. This research project could easily be used to obtain an idea of user requirements for enhancing the existing system. What types of salesmen under what conditions are likely to be heavy users of an information system? What are their needs when compared with salesmen who do not use the system? This is an inexact approach to design, but systems design procedures have to be developed to fit each situation.

In addition to looking at descriptive models of individual decisionmakers, it is possible to apply some of the techniques presented in Chapter 2 to determine user needs. Many interesting data come from open-ended questions on a general user survey. Another technique is to distribute copies of existing reports and ask for comments on the data currently in use. The

user might be asked to speculate on what data could be added
to the report to make it more useful for him. Such a procedure
is good practice after a new system has been implemented in
order to determine how it is being used and what changes are
needed.

TRAINING AND UNDERSTANDING

In some of our user surveys, training has frequently been
mentioned on questionnaires as a problem. Users have expressed
the desire both for more general training on information systems
and computers and for specific training on how to use existing
systems. One manager of information services whose company
made heavy use of a file management package with users
actually completing the input forms admitted that he had never
spent anything on training!

The first type of training is a basic one and applies to all
systems. What inputs will be provided? How should they be
filled out? What should be done with output reports? Sufficient
documentation should also be available to answer questions when
someone from the information services department is not available. A good example of the kind of complete user documentation needed in order to use a system can be found in the write-up
of the *Statistical Package for the Social Sciences* (Nie et al.,
1970). This particular program package is designed for use by
a number of different researchers. The book contains many
examples of how the package is to be used and is very clearly
written; it can serve as a guide to writers documenting systems
for users. (Complete guidelines on nonuser documentation can
be found in Harper (1973).)

In addition to training for the mechanical operation of the
systems, it is important to train users to understand the system.
Ackoff (1967) has described a situation which counters the
argument that a manager does not have to understand a system
in order to use it. He describes an inventory control system in

which plans were being made to add to the current $2,000,000 investment in the system. In questioning the operation of the present system, Ackoff found many parts in the inventory at their maximum quantity despite the use of an economic order quantity model. Upon careful examination it was found that the system was reordering when the maximum amount was reached. The economic order point was being confused with the maximum allowed in inventory. These and other errors produced an estimated system cost of $150,000 per month over the manual system it replaced, mostly through excess inventories. Neither management nor anyone else in the firm understood what the system was doing. This and other examples demonstrate the importance of users understanding system processing.

The analyst and decisionmaker must work together to develop descriptive models which help in understanding system requirements. Iterative design procedures in which the user develops his own reports will also help. Another reason for having the user design as much of the processing logic as possible is to increase his understanding of the new system.

ACCURACY

Accurate information systems are important, particularly when they deal with financial and accounting information. Users often view problems in accuracy as transcription or keypunch errors. Because of inadequate training, users frequently have no conception of files or file maintenance. Bad input over a period of time begins to erode the integrity of a file, but users do not realize the magnitude of this problem.

Accuracy problems were also raised in the discussion of the physical user interface. The quality of systems depends partially on being able to process the data accurately and on schedule. Are the original sources of data accurate? Is there incentive to provide the needed input? The framework may be useful here because for certain types of decisions complete accuracy is unnecessary, such as in the strategic planning area. Accuracy should be treated as a design parameter; there is no need for the user to pay for extreme accuracy if it is not needed. However, it is important for these design decisions to be made jointly with the users rather than by the systems staff. It is also a good idea to follow up with post-design audits to be sure that accuracy is maintained during the operation of the system. Sampling techniques can often be used to reduce the processing requirements of these audits.

DEPENDABILITY

Dependability means many things. For a batch system, is the output provided on time as scheduled? While the system may be designed so that processing can be done, it must be executed on a computer system which is loaded with other applications. Can this new application be added to the existing schedule and dependable operations maintained? While these conditions are not strictly the fault of the systems designers, the user does not realize that the system is undependable because of production problems in the computer center.

For on-line systems, dependability frequently means availability. Failures or down time creates a real problem when the user has become dependent on an on-line terminal. Typically these systems deal with operational control decisions which are the most critical in the day-to-day operations of the organization. On-line systems must provide adequate backup and recovery, as users quickly become dependent on them and are highly critical when the system is unavailable.

USER REPRESENTATIVES

The use of a computer ombudsman or a representative has been suggested by Lucus (1973a) and in the last chapter. This individual's duties begin with systems design; he helps to integrate the user department and the information services department. During the operation of the system he is the key contact point for each application if something goes wrong. This representative monitors errors and the accuracy of the system and suggests enhancements which should be made as a continuing part of systems design. The average organization may need several of these individuals, each assigned to a different series of applications or departments.

FLEXIBILITY

One of the most frequent complaints from user surveys is the lack of flexibility in systems. Each user has his own needs and desires; instead of meeting these needs, we have tended to build rigid systems which force users into a common mold. Flexibility applies to two different aspects of information systems. First there is flexibility in designing the system so that it provides the user with the information he needs. The second type of flexibility is in making changes in systems as new needs arise. Unfortunately, there is no single firm answer to providing flexibility in systems. There are many different situations, and we shall discuss some approaches that can be used to increase the flexibility of information systems.

ON-LINE SYSTEMS

In fully on-line systems in which updating is done on-line, flexibility for the user to fulfill his needs is almost always inherent in the design of the system. The selective retrieval of information is generally possible in on-line systems. Usually the input is also flexible in such a system in the sense of giving immediate feedback and explanations of what is wrong. Users can enter data in a flexible format using key words or codes.

The second type of flexibility, that of changing a system as new needs arise, is very difficult with an on-line system. These systems have complex direct-access file structures; there may be many pointers and directories, so that adding fields or changing files is a difficult process. Major reprogramming and conversion steps may be required. The cost of developing and programming a fully on-line updating system is usually high and the development period is lengthy. Aside from some well-publicized examples, few of these systems exist today. Instead, most users are confronted with batch processing systems.

BATCH SYSTEMS

Batch computer systems are generally considered to be inflexible; they are associated with strict deadlines for input and for processing, along with rigid formatting requirements for input; output is often inflexible and consists of preformatted reports. Fortunately, such systems are more flexible than on-line systems with respect to changes; file structures are usually less complex, as are programming requirements.

There are some simple methods for developing flexibility in batch systems. First, exception reporting is highly recommended; data are provided only when information exceeds historically derived ranges specified by the user. Users will probably have to be trained to accept this form of output; however, Pounds (1969) found that many managers used historical models in decisionmaking. An information system can serve to stimulate the intelligence phase of decisionmaking when deviations from past history are noted. As long as conditions are operating according to these historical models, there is no need to alert the user. Clearly, such a radical departure from the principle of full printout requires working with the decisionmaker to convince him of the benefits of this approach. Information may often be looked upon as security or as a competitive edge in the organization, even though it is not used. Users may be reluctant to sacrifice full and completely detailed reports and substitute more streamlined exception reports.

The next step beyond exception reporting for more flexible output is to provide some type of a data management and retrieval package for the decisionmaker. Either the manager can select several standard formats or possibly a report generator can be used for retrieval. For example, there are a number of popular data management systems which allow the user to formulate his own retrieval requests.

AN EXAMPLE

As an example of providing flexibility in batch systems, consider the following sales information system serving over 400 salesmen and 100,000 accounts. The system supports a variety of decisions and serves salesmen and managers. Some of the salesmen have 80 to 100 active accounts; others have only 1. Currently, all salesmen receive virtually the same output; the complete report which requires 37,000 pages of printing a month for all users is divided among salesmen and managers.

From a user questionnaire survey it was found that some of the salesmen wanted data in more detail, such as by item or color and style, and others wanted more summary information than the basic product line data currently provided. It is probably even true that some of the salesmen want different details for different accounts. How can all of these conflicting needs still be met with a batch system?

One approach to increase the flexibility of the system would be to build a file of salesmen that contains a short code to indicate the type of report each salesman wants. During processing, report format and appropriate level of detail could be selected on the basis of this code. In a sense, each salesman would choose from among a series of different report options available. This facility would require a more complex program plus a small amount of file storage, but the requirements would not be too severe. Additional flexibility could be provided with a direct access file containing the salesman's desired output formats by account. Each man could have a different format for each account.

The next step in providing flexibility might be to make the salesman output file available on-line during portions of the day for changes. The system could be extended even further by letting the salesmen enter accounts on-line whose reports he wishes to see during the next week; this capability would reduce total printing and spread the processing load more evenly.

Finally, it would be possible to move to on-line inquiry of the entire data base. This would be an expensive option because of the large size of the master file. However, storage costs are decreasing, and it might even be possible to regionalize the data base and have only certain portions available on a schedule which takes advantage of different time zones across the country.

All the suggestions above have been alternatives to a fully on-line, updating system. Many times it is not necessary to update a file on-line, but it is desired to reduce the amount of output produced. Techniques such as those above can be used to reduce output and to provide more flexibility in meeting user needs with varying levels of complexity and cost.

For any system, either on-line or batch, design should be as modular as possible and should reflect the expectation of future changes. The desire to make changes is one of the most important reasons for documenting systems and programs well. If information systems are to be useful to a large number of decisionmakers, they have to provide flexibility. The flexibility of giving the user what he needs has to be built into the design of the system. Flexibility also means that it is possible to change a system after it has been implemented. While no programmer or systems designer likes to modify a working program or system, change to existing systems should be expected.

FUTURE SYSTEMS AND FLEXIBILITY

The maximum amount of flexibility in a system can be found in the examples of decision systems developed to support a single decisionmaker or a small group of them. According to the Gorry–Scott Morton framework, we expect to see many more of those systems developed in the future. (One of the systems is discussed in detail in the next chapter.) These systems are designed with and for a particular group and require an intensive development effort. The decision situation is studied in detail and a rough version of the system is developed; several iterations are required to modify the design. Experiments are conducted in which the user works with the system and suggests changes to it. These systems represent the most flexible information system in existence at present; their primary drawback has been their high cost. Developing one of these systems is a labor-intensive process and requires a long period of time.

One alternative to such a system would be to use a simple time-sharing computer with a low cost, storage tube graphics CRT for output. A higher level time-sharing language and an existing operating system can be used to handle the major programming burdens; the staff has no telecommunications processing or systems programs to write. System design with such a low-cost alternative still requires a labor-intensive effort to build a decision model, but it should be relatively cheap and easy to implement the system. With an inexpensive time-sharing system and output device, it would almost be cost effective to support one-shot decisions and throw away the system when done. It would be possible to develop such a system with external time-sharing service bureaus. Summarized data could be transferred to the system and on-line programs could be used to manipulate the data. For decision support systems, on-line updating is usually not considered as important as on-line access and interaction (Scott Morton, 1971).

SUMMARY

The quality of systems as defined by users includes the technical and organizational considerations summarized in Table 5.1 as a minimum. Evaluating systems based on user-defined quality is central to our philosophy of creative systems design. This approach to systems quality encourages user participation in the design process, reduces conflict, and lessens the amount of power transferred from users to the information services department. The preceding

TABLE 5.1. CONSIDERATIONS IN
DESIGNING HIGH QUALITY SYSTEMS

Quality defined by users

Goals
   Established jointly
   System in best interest of user and organization

Proper functioning
   Manual procedures
   User interface
   File maintenance

Design alternatives

Design Considerations
   Decision analysis
   Filtered information
   Descriptive models of decisions

Training
   Operations
   Understanding

Accuracy

Dependability

Flexibility

---

chapters have offered a number of suggestions for improving systems design through a creative, user-oriented approach. The next chapter discusses three very different examples of creative design to demonstrate how some of these ideas have actually been implemented.

RECOMMENDED READINGS

Ackoff, R. L. 1967. "Management Misinformation Systems," *Management Science*, 14 (4):B147–56. (A classic article pointing out several major systems design assumptions which are often incorrect.)

Driver, N. J. and S. Streufert. 1969. "Integrative Complexity: An Approach to Individuals and Groups as Information Processing Systems," *Administrative Science Quarterly*, 14 (2):272–85. (A somewhat difficult article discussing the theory of integrative complexity and information overload for individual information processing.)

Lucas, H. C., Jr. 1973. *Computer Based Information Systems in Organizations*. Palo Alto, California: Science Research Associates. (See the section on Systems Analysis and Design for some general considerations in developing high quality systems.)

Pounds, W. F. 1969. "The Process of Problem Finding," *Industrial Management Review*, 11 (1):1–20. (An extremely insightful article describing managerial decisionmaking and the importance of historical models in problem finding.)

CHAPTER SIX

THREE EXAMPLES

INTRODUCTION

THE PURPOSE of this chapter is to illustrate some of the techniques discussed earlier and to show how they can be combined into an approach to systems design in different settings. Creative systems design, however, is not one single solution to every problem; it is an approach which leads to actions appropriate to each situation. First, the design team accepts the idea that systems design is a planned change process. The team recognizes that the major problems in developing a system are likely to be in the organizational behavior area. The design process is carefully planned to encourage the user to participate and to define the system's requirements. Attention is given to developing a successful user interface and the quality of the system is defined from the user's perspective.

Three applications are presented in this chapter; the first is an information system for the United Farm Workers Organizing Committee which involved the author and a group of students.

The development of the second system was guided by a Ph.D. student and is an information system for a tax shelter, cattle investment firm with no prior computer experience. Professor Michael Scott Morton of MIT designed the last system, which is illustrative of a dedicated, decision support system.

The purpose of these examples is simply to illustrate how the various techniques discussed in earlier chapters can be applied. The fact that these systems seem to be successful on a number of dimensions certainly does not prove the efficacy of the proposed techniques since we have only three cases to consider. However, because we lack sufficient knowledge at present to develop a "formula" for creative design, these examples illustrate how the appropriate techniques can be used in each different circumstance.

The research nature of the systems described in the rest of the chapter should also be noted. The participating organizations were probably more open to the use of new techniques because of the research environment. The researchers were also more interested in trying new approaches than they might have been in a nonresearch setting. At the same time an attempt was made in actually working with the organizations to prevent the research from interfering with the design of the system. Care was taken to produce a useful system and to explain the importance and relevance of research activities in developing the final system.

UNITED FARM WORKERS ORGANIZING COMMITTEE

BACKGROUND

In the summer of 1970 a representative of the United Farm Workers Organizing Committee asked the Graduate School of Business (GSB) at Stanford to discuss ways to improve information processing in the Union. The Union had met recently with computer manufacturers and service bureaus for help; on the basis of this experience, Union representatives felt the need for more objective viewpoints. We visited the Union to discuss their needs and the form of a possible relationship between the Union and the Business School.

The Farm Workers Union was established to secure better wages and working conditions for farm workers, many of whom are members of minority groups. The Union is headed by Cesar Chavez and is staffed by volunteers who receive a small weekly allowance plus the expenses necessary for subsistence. A major goal of the Union has been the elimination of migrant labor contractors and the substitution of a Union-operated hiring hall. The Robert F. Kennedy (RFK) Memorial Insurance Plan pays certain medical costs for workers and their dependents. Eligibility for these benefits is based on the amount of time worked by the union member. The insurance plan is a trusteed fund supported by grower contributions and administered by a joint board of Union and grower trustees.

At the time of our initial visit, the Union had just signed contracts with major grape growers and a large increase in membership was expected. One of the Union's major information processing problems is keeping track of membership records, a vital function for providing Union services to members. The Union also has to record the hours worked by each member from data supplied by the growers in order to determine RFK Plan eligibility and level of benefits.

APPROACH

On the basis of the first meeting, a joint research project between the Graduate School of Business at Stanford and the Union was established in order to develop a system. We would conduct research on the process of systems design and would be able to involve students in field settings. The first objective was to analyze present information processing capabilities and determine if a computer-based system was feasible. In return for the research opportunity and student involvement, the Union would receive assistance from the design team.[1]

The arrangement was particularly crucial, and it represents the development of a "psychological contract." It was clear that the design team could only perform a feasibility study and the Union would decide whether to implement the system. The design team from Stanford would not program or implement the system; the Union would have to obtain such services or hopefully develop the internal capability to implement any system itself.

The design team was eager to work toward the development of a relationship with the Union based on trust so that worthwhile recommendations would be accepted. A contact in the Union who would be concerned full-time with information processing was requested by the design team. The major question facing the design team was whether or not a system could be developed which the Union could implement without dependence on an external group. A second concern was the impact of a computer system on the Union, a flexible, dynamic, action-oriented organization. What would be the impact of a rigid computer system on such an organization?

The design team worked to develop a strategy which considered these concerns. First, an effort was made to maintain a contact within the Union who would work full time on data processing activities. If the system proved feasible, the second major component of the strategy was to design as simple a computer system as possible. In terms of the framework of the first chapter, the system would focus on processing transactions and the support of operational control decisions. A simple system would be the easiest to implement, would have the most straightforward user interface, and would have the greatest potential benefit for the Union. A more comprehensive system would require a greater development effort and would be more likely to en-

[1] The research was supported in part by a grant from the Donaldson, Lufkin, Jenrette Foundation.

counter difficulties. It is always possible to expand a system later to encompass more activities, but the design team felt it important for the initial effort to be successful.

To learn about the potential impact of a computer system on the Union and to monitor the impact of the systems design effort, the techniques suggested by Lucas (1971a) and in Chapter 3 were adopted. A questionnaire was administered to Union staff members to assess their attitudes toward the computer system and to obtain data for designing a system. For example, what departments in the Union were in greatest need of assistance in handling their information processing responsibilities? The design team also stressed participation in the feasibility study and the subsequent design of the system; review meetings were held with potential users and open communications were encouraged.

DESIGN

The questionnaire analysis provided helpful information both for design purposes and for assessment of the readiness of the Union for computer processing. The design data demonstrated a commonly held set of priorities; all Union members agreed that the membership office and the Robert F. Kennedy Health Plan were in most need of assistance. The attitudinal data showed very high expectations about computer systems on the part of all staff members. Unfortunately, the least favorable attitudes were found in the Robert F. Kennedy Plan office and the membership office. These findings had to be carefully considered in approaching the area, and the design team tried to make Union leaders aware of the potential problems in these two offices.

The attitudinal data also showed that Union leaders were more enthusiastic about the contributions from a computer system than were other staff members. The design team alerted the Union leaders to this problem so they would know that their optimism and expectations were not necessarily universally shared. The Union leaders needed to help prepare the staff to accept the system and cooperate in its design. Without the information furnished by the questionnaire, Union leaders might have felt that the rest of the organization shared their eagerness for a system. Instead, provided with this information, the leaders were able to help assuage the fears of some of the other staff members.

Student volunteers gathered data for design and by January 1971 the Union and GSB design team concluded that a computer-based processing system was both feasible and desirable. After a change in the Union contact and some uncertainty about the relationship between the Union and the design team, a full-time Union member assumed the responsibility for information processing within the Union. The GSB design team shared all of the data from the design system to date with this contact and, in retrospect, this individual is clearly responsible for much of the later success of the systems design effort.

During the early part of 1971, a rough draft of the proposed system was developed. Staff members attended a review meeting at a Union retreat to discuss the preliminary design. Cesar Chavez began the day-long session by saying that no one would be replaced by any computer system. There was much more interesting and important work available within the Union. Instead of concentrating on the impact of a system on their jobs, he requested the staff members to think about the functions to be performed by the system and to concentrate on improving the system from the overall standpoint of the Union. The design team began by announcing that the system proposed was a rough draft and that the meeting would be a success only if modifications were made to the system. A brief tutorial on computers and information systems in general followed this disclaimer. There was a casual presentation of the proposed system; flip charts and elaborate view graphs were absent. The design team described the system from lecture notes and prepared charts in front of the audience.

The presentation began with the file contents and then progressed to reports, finishing with the description of input documents. The audience responded extremely well to the presentation; staff members seemed to understand the system and offered many suggestions for changes. Though the group assembled had a low degree of formal education, there was high motivation. Because of the intense interest and the level of understanding of the Union staff, major portions of the system were redesigned in response to the suggestions. Many of these changes were incorporated on-line in front of the audience, thus reinforcing the impact of the suggestions.

Following this review meeting, Union leaders participated in several subsequent meetings after they had the opportunity to think further about this system in order to suggest additional managerial reports. The design team helped to develop bid specifications which were submitted to outside vendors. The cost of outside services was too high, and the Union decided to develop its own internal capability to finish the design and implement the system. Several highly qualified programmers joined the volunteer staff of the Union and a firm donated computer time for debugging and implementing the system.

A follow-up questionnaire was sent to the Union staff in the spring of 1971 to assess the impact of the design on the Union and to prepare for conversion and implementation. The follow-up study showed that, in general, the response of the Union to the idea of a computer system continued to be favorable. The only disconcerting finding was that of a perceived low level of involvement on the part of staff members in the design. This feeling existed despite a series of meetings and the design team's emphasis on participation. The GSB team discussed the results with the Union computer staff and encouraged them to stress user involvement and review as implementation progressed.

RESULTS

The system became operational in January 1972, and the design team from Stanford supplied only occasional advice and participated in a few meetings during the implementation. The system is viewed as a major success. The staff, from Cesar Chavez through clerical personnel, are visibly enthusiastic about computer processing. The computer system will be expanded in the future, and users have suggested many additional subsystems. It appears likely at this point that the Union will obtain its own computer system. The major attention to the organizational problems and insistence upon keeping the first system simple were major factors in this success. The successful results of this first effort with computers have given the Union the internal expertise and confidence it needs to continue developing useful information systems. For more details on the development of this system and the consulting relationship between the Union and Stanford, see Lucas and Plimpton (1972).

A LIVESTOCK MANAGEMENT SYSTEM

BACKGROUND

The organization involved in this systems design effort is a small investment company in Southern California. The company is responsible for maintaining client investments in livestock for breeding and marketing. Four or five managers are involved in the livestock division, including the president of the firm, a vice president for livestock, an accountant, an operations manager, and two field managers who had not yet been hired at the time of the study. The investment firm has contracts with a number of ranchers to provide services and employees (ranch hands) to manage the livestock. In the spring of 1972, the company managed some 5000 Angus cross-bred heifers and some 5000 calves which arrived during the project.

Two students undertook this research project as a part of a class project; however, one of these students, a Ph.D. candidate in Food Research, continued the project beyond normal course work requirements because of his interest in the subject. The investment firm was more committed to change and to a different form of management than other companies in the industry. The objective of a computer system was to handle tax reporting, inventory management, herd management, and planning (for example, for breeding). The personnel in the organization were young and had relatively little experience in the livestock field; they worked easily with the students and developed a good relationship with them.

APPROACH

The systems design team knew little about the operations of this part of the cattle industry. They held a long meeting with livestock division personnel, in which the systems designers learned about the industry and its problems. The designers tried to determine the perceived needs of each user. The design team spoke of other computer applications in general terms; they did not try to claim how a specific system might help the company at this point. After discussing potential problems with the entire management group, the design team began to sketch how a system might be built to handle the total set of problems.

In order to maintain as much flexibility and objectivity as possible, the designers completed the design before choosing a language or the type of processing (batch, time sharing, on-line, etc.). The students saw that this was a flexible organization, and they included features to keep the openness and climate intact after the implementation of the system. The design team developed four specific innovative procedures for this situation. First, the word "computer" was not used at all during the initial discussions and was used as little as possible in later meetings. As the designer spoke of other applications, the users began to ask whether similar things could be done for them. The designer did not try to force a computer or particular system on the organization.

The designers developed the second innovative procedure because of the distance between the design team and the company; mail requests were often unanswered. The designers wished to have the users develop as much of the system as possible. They introduced users to flow charting and asked them to develop the major logic of the system. In order to work successfully, given the distances involved, the designers divided requests into parts and mailed them to users. If there were two parts to be developed, two users were each sent two quarters and requested to do one quarter and give the other quarter to the other user. Thus each user completed one half of the request: his own quarter and a quarter from the other user. The whole project seemed smaller, and a situation of mutual dependence was created. Each user of the pair had to help the other; this system avoided requests being given a low priority and being ignored for a long period of time. This approach of dividing a single task into multiple parts and asking one user to be responsible for seeing that the other parts were done was labeled "Balanced Work Request" by the design team. The results from this technique were impressive as response time went from nothing to an average of a four-day turn around for answers to the requests.

The designers developed "Creative Conflict" as the third major component of the design strategy. Each staff member created data on two imaginary head of cattle; the only restrictions were that each animal had to be logically possible according to the fields defined for the data base. Descriptions of each animal contained data for all the major parts of the system; this device forced each member to learn something about the other operations. One imaginary head of cattle was to be the ideal one for each user department's needs. For example, the tax man would design an animal which optimized tax advantages and resulted in the biggest gain for investors. The operations manager might design an animal to gain the most weight by sale time and have the least cost per pound. The second animal developed by each user had impossible data conditions; here the user was asked to force another department's processing algorithms to fail. The president of the company established a prize, and each man tried to design his processing to be foolproof when confronted with the animals with impossible data. In this instance, the design team converted what could have been a situation of potential conflict into a challenge. The participants perceived the game of creating animals and designing one's own processing as fun.

Finally, the design team spent a minimum of one-half hour discussing the objectives of each coming half-day work session. Usually the discussion was held away from the office during a meal. With these objectives clearly defined, little time was wasted during actual meetings.

DESIGN

The strategy described above essentially resulted in each user designing his own portion of the system. From the decisions described by the managers, the systems designers developed data fields and defined their size. The designers merged the fields together for all of the users to eliminate duplication, and a variable name for a programming language was provided for each field. The flow charting was an iterative process because complex calculations were involved, for example, sorting on the genetic qualities and developing breeding probabilities, or performing tax computations. The systems designers worked closely enough with users that users actually defined the logic using the variable names assigned to data fields. Soon the users were drawing and modifying flow charts and filling in the details. Users also designed their reports on printed grid work sheets.

The designers have built a macroprogram which calls each user's subroutines. Because of cost and the fact that some data (for example, on breeding) will not be available for several years, the staff plans to implement the system gradually.

RESULTS

While the system is not yet fully implemented, portions are working. Probably the most important feature of this system is that users actually designed their own system. The designers facilitated the users' enunciation of their own needs. The system embodies fairly complex logical processing, yet through creative design techniques all users ended up cooperating in the design. The potential for destructive conflict was avoided entirely or minimized. As a result, the managers understand the whole system since they participated in the total design. Instead of being an onerous chore, the systems design approach resulted in a learning experience which will be beneficial to the users when the system becomes operational.

A MANAGEMENT DECISION SYSTEM

BACKGROUND

The two preceding systems process a large number of transactions in batch mode and can be classified as transactions processing and operational control decision systems according to the modified Anthony framework of Chapter 1. The system described in this section, developed by Professor Michael Scott Morton at MIT, can best be classified as a decision support system falling into the managerial control decision category of our framework (Scott Morton, 1967, 1971). The setting for this system is a major manufacturing company with some 70 divisions making products ranging from electric toothbrushes to industrial turbines. All operations are decentralized into divisions which operate on a profit center basis. The division involved in this particular study sells laundry equipment (washers and dryers). The system developed supports decisions for three managers in this division. The marketing manager is primarily interested in sales, and the production manager wishes to minimize production costs and inventory levels. The marketing-planning manager has the responsibility for planning and integrating the diverse goals of the other managers. Monthly, a production and sales plan for the next twelve months is developed for the manufacture, sale, and distribution of laundry products. A number of variables are included, such as determining the amount of promotion and advertising, scheduling production, purchasing, planning work force levels. The major external variables are inventory levels and the type of customer service provided. The potential users understood that the system was being developed as a research project to help improve the decision-making process.

APPROACH

The systems design group participated in meetings with the managers for some six months; there were three months of intensive observations and three months of spot checking. Trips were made with managers to the factory, warehouse, headquarters, and other locations to attend further meetings involved in the planning process. The designers built a descriptive model of the decision-making process and worked with managers to identify their objectives. All of these results helped to develop a normative model which was contrasted with the descriptive model to identify bottlenecks in the planning process. A system was developed to deal with the bottlenecks and remove these constraints on the planning process.

DESIGN

After four months of observations, work began on developing a system which featured visual, interactive graphics displays as the primary user interface for input and output. In order to avoid raising expectations while the designers continued to observe the decision process, the design team did not discuss various alternatives with managers. After debugging the nucleus of the system, the marketing-planning manager and his staff worked with the device and he agreed to act as a liaison with the other two managers.

The designers undertook two months of training the marketing-planning manager using test data; the training sessions resulted in numerous modifications to the system. Teaching the use of the system was a challenge; the manager learned the individual control points and how to operate the light pen on the CRT easily. However, he had difficulty combining the techniques into more complex and meaningful patterns of action. Experience with the tests suggested many system changes in the user interface which reflected the habits and styles of the marketing-planning manager. After this initial period of modification, the marketing-planning manager taught the other managers how to use the system. Even though significant improvements had been made by this time, the system was changed further as required by the reactions of the other managers. The designers reported that the system was modified continuously during use. They obtained data on user reaction through the use of a tape recorder and a Polaroid camera. When analysis of these observations showed an opportunity for improvements, the designers reviewed changes with the managers and implemented them if desirable.

RESULTS

The system was used heavily by the managers and became an integral part of the planning process. The managers clearly preferred the system to previous methods and felt that they made better decisions in less time. The heavy participation in design was partially responsible for the regular use of the system. Also, by studying user needs and being responsive to suggested changes, the designers developed a useful system which became an essential part of the planning process. Attention was paid to maintaining the usefulness of the system through modifications.

The designers reported that the time spent by the managers in the planning process dropped from an elapsed period of three weeks to roughly half a day, and the managerial time consumed by meetings dropped from six days to one-half day. One major advantage was the continuity created by the system. No session had to be interrupted to give output to a group of clerks who performed massive computations before the next planning session. Some of the time reduction and continuity in decisionmaking came from an interactive system as opposed to a manual or batch processing system. The managers could test alternatives and see patterns graphically. In contrast to the manual method, managers did not accept the first feasible solution but explored a number of feasible solutions. The system also enhanced communications among the decisionmakers; they developed a clearer understanding of the decision process and a greater commitment to the plan.

SUMMARY

LIMITATIONS

The examples described in this chapter represent a wide range of applications, from simple batch business systems which are primarily transactional in nature to interactive graphics used in a decision support system. In each instance the designers recognized the planned change nature of information systems development and used appropriate techniques to deal with the situation.

The systems were developed under somewhat unusual circumstances. Students or academicians working on research projects designed each system. Diverse organizations participated in the projects, though a relatively small number of individuals were involved, ranging from thirty in the Farm Workers project to three in Scott Morton's management decision system. There was a high probability of failure in all three instances because of the use of outsiders to design the system. Had these external change agents attempted the force a system on users or failed to explain their motives, user resistance might have been high. On the other hand, the research setting made all parties more open to experimentation, thus partially offsetting the problem of the designers being "outside researchers."

REVIEW OF TECHNIQUES

All three examples demonstrate the attitude and approach to systems design stressed in earlier chapters. Attention focused on designing a system appropriate to user and organization needs in terms of technology, response time, cost, and probability of success. The designers used a variety of techniques to elicit user participation, including formal questionnaires, observation of problem-solving sessions, and system review meetings. The designers attempted to have the user either develop the system initially or modify it heavily. The atmosphere encouraged change; designers did not defend a vested interest in an already frozen systems design. In two of the instances, the user participated heavily in the actual design of system logic. In the other situation, the users constructed the flow charts and developed files. The designers paid careful attention to the way in which the user interfaced with the system. Finally, the design teams conducted extensive experimentation with the systems and made modifications to fit used needs. A summary of the techniques used during the process of system design, developing the physical user interface, and defining the quality of the system may be found in Table 6.1.

It is hoped that these three examples have demonstrated how different creative approaches to system design can be integrated and applied to specific situations. Not all techniques were appropriate in each instance, but each approach included cre-

TABLE 6.1. CHARACTERISTICS OF THREE DESIGN EXAMPLES

Component	Farm Workers Union	Cattle Investment Co.	Management Decision System
Process of design			
Planning	Research project; Impact on organization, work groups stressed; GSB team only aids in design, Union implements	Users to design system; Creative conflict to minimize group conflict; Balanced work requests	Research project; Goal to build descriptive model of decision process; Plans to work closely with managers
Change program			
Participation	Attitude surveys to assess impact, plan design priorities; User representative in charge; Review meeting emphasizing change; Changes made as suggested by users	Users designed files; Users designed processing algorithms; Users drew flowcharts	System taught to liaison manager ("county agent"); Liaison taught other users; Changes made willingly; Continuous changes
Data collection	Questionnaire surveys on attitudes and design; Interviews	Interviews	Interviews and observations
Physical user interface			
Type of system	Batch	Batch	On-line
Input	Form design reviewed with user	User designed forms	Input designed to fit user requirements on CRT with light pen; New commands added as needed
Output	Clear headings; Spanish and English where needed; Full error messages	User designed reports	Graphics CRT with light pen; Researcher designed, modified by user requests; Filtered, relevant data
General	Reviewed with user; Minimal duplication; Well-designed manual procedures	Minimal duplication; Few manual procedures; Phased conversion and implementation	Elimination of previous manual operations; Testing with user leads to modifications

Quality of system	Thorough testing Planned conversion		
Definition	Quality defined by user indirectly	Quality defined by user directly	Quality defined by user through observation
Goals	Goals agreed upon with users from questionnaire, meetings	Goals defined by users	Need expressed by organization
Proper functioning	Simple system—primarily transactions oriented	Transaction, operational control, some planning	Technically complex system, managerial control application Design iterations to assure proper operations
Design alternatives	Critical areas included in system Did not select one computer system for processing Alternatives suggested by users included if possible	Language and computer not selected until design completed Alternatives selected by user for his area	Researchers designed basic system, selecting CRT system Alternatives suggested by users incorporated in system
Design considerations	Filtered output Descriptive models developed from questionnaire, meetings	User designed to fit his own decision model	Descriptive model of decision compared with normative to obtain design
Training	Careful training so users understood system Review sessions inaugurated training	User designed, so understands own area Creative conflict technique helps user learn about other areas	Liaison manager taught system Modified to fit his style Liaison trains other managers System modified to fit new users' needs
Accuracy	Highly accurate—large improvement over manual system	More accurate than a manual system	Accurate as needed
Dependability	Dependable batch processing with easy schedules	Dependable batch processing with easy schedules	Dependable equipment System available as needed
Flexibility	Flexibility emphasized in design; less flexibility once implemented Changes made as requested	Flexibility emphasized in design Some flexibility for data analysis after implementation	Flexible input/output Selective retrieval Flexibility in making numerous changes after system in use

ative system design techniques appropriate to the environment and situation. The next two chapters deal with topics closely related to creative systems design, mechanical aids to systems design and project management.

RECOMMENDED READINGS

Gerrity, T. 1971. "Design of Man Machine Decision Systems: An Application to Portfolio Management," *Sloan Management Review*, 12 (2):59–75. (This article discusses the design of an on-line, CRT-based information system for bank portfolio managers and presents some statistics on its use.)

Lucas, H. C., Jr. and R. B. Plimpton. 1972. "Technological Consulting in a Grass Roots, Action Oriented Organization," *Sloan Management Review*, 14 (1):17–36. (A historical discussion of the Farm Workers project with emphasis on the relationship between the consultants and clients.)

Scott Morton, M. S. 1967. "Interactive Visual Display Systems and Management Problem Solving," *Industrial Management Review*, 9 (11):69–81. (A summary of the major design consideration and results from implementing the system described in this chapter.)

——. 1971. *Management Decision Systems*. Boston: Division of Research, Graduate School of Business Administration, Harvard University. (A complete description of the research design and results for the system described in this chapter.)

PART THREE

SUGGESTIONS FOR SOLVING
SOME TECHNICAL AND
MANAGEMENT PROBLEMS

CHAPTER SEVEN

MECHANICAL AIDS
TO SYSTEMS DESIGN

INTRODUCTION

IN THE PRECEDING CHAPTERS we have focused on problems and solutions to organizational behavior problems in systems design. The first priority is for the design staff to recognize these problems and plan for their solution. Having adopted a creative approach to design, the staff can profit from a number of mechanical aids. These aids save development time and introduce additional flexibility into the design process to improve the overall quality of systems. Mechanical aids to design can contribute substantially to the creative design process.

In this chapter we look at existing aids and software packages. Most of these products represent an attempt to solve one small problem in systems development; they do not form a coordinated approach to the problems of systems design. We also examine two comprehensive systems under development or in the prototype stage which provide a more unified approach to augmenting the systems design process. One of these systems in particular has the potential for revolutionizing many of the technical aspects of systems design, allowing designers to focus more on behavioral and managerial problems.

APPLICATIONS PACKAGES

TYPES OF PACKAGES

File and data base management packages have been some of the most successful ones in the computer industry (Byrnes and Steig, 1969). File management packages are frequently used to process inquiries against a file created by a system developed by the information services department, though many of the file management systems include updating capabilities too. These systems have evolved over a number of years and are in widespread use. Generalized data base management packages are more comprehensive; they usually are integrated with the operating system and provide support for a number of activities including data base creation, accessing, auditing, editing, protection, updating, and teleprocessing (Sibley and Merten, 1973).

The basic purpose of any package is to extend the capabilities of a computer system to the end user. Most of the file management packages use either a free-form language or a highly formatted language for input. There is some question, however, whether the packages have really extended the computer to the end user because of the complexity of the request procedures. In one Bay Area company, user departments did submit requests using a file management package, but the requests were filled out by a systems specialist in each department rather than by the users themselves.

In addition to the general file and data base management packages there are available a number of special purpose packages such as payroll programs, piece parts explosions, and inventory control systems. These are aimed at a specific function, for example, retail and wholesale inventory operations. The goal of the company developing the package is to find some application which is general enough that it will fit a wide variety of organizations, but specific enough that it will be reasonably easy to use. When such conditions are met, a package can be much cheaper and more quickly implemented than an application developed from scratch.

Another type of package is one in which the entire computer system is dedicated to the application. The best example of such a program is PARS, the IBM Passenger Airline Reservation System. This particular package is built around the nucleus of the operating system OS360/370 and is based on IBM's experience with the development of American Airline's SABRE system. The package is in use by a number of airlines today, and it provides the basic logic for making reservations. It supports all the tele-communications and operating systems functions of an on-line passenger reservation system. Each airline using the system has to provide parameter tables which describe its route structure and special conditions. Some of the airlines using the package have modified it to add different capabilities and to customize the system to their needs.

A dedicated package such as PARS is similar to any special purpose application package, but the effort involved in implementation is much greater. Given the magnitude of the task of developing code like the control programs for an on-line reservation system, using such a program as a nucleus is clearly easier than building the entire application. In fact, two airlines tried to develop their own packages with the aid of different manufacturers; these efforts ended in law suits between the airlines and the computer vendors, each claiming breach of contract. At least one of the airlines involved abandoned the effort to develop its own system and adopted the PARS package.

The last category of packages includes mathematics and statistical programs which have been used for a number of years in the academic environment and in industry. These packages include BMD, SPSS, and IBM Fortran library SSP, and many other special packages. The programs are very problem oriented; the BMD and SPSS packages are completely self-contained, whereas the Scientific Subroutine Package is embedded in a host language. In this situation where a number of users are performing almost identical analyses of data, it is clearly easier to use a package than to write individual programs.

DECISIONS ON PACKAGES

The major issue for both the designer and the user of an applications package is the tradeoff between efficiency and generality. One goal of the developer of the package is to make it sufficiently general that it can be applied in a number of different situations. In conflict with this goal are the objectives of efficiency and ease of use. If the package is too general, it will be very inefficient in execution and frequently will have more capabilities than each individual user needs. This means that the user must supply a number of parameters which really have no relation to his particular needs. The file management packages which have probably been the most successful can be classified as general purpose. However, they are clearly focused on the task of file management and data retrieval; the packages contain a defined set of functions and differ primarily in the functions included and the form of the input and output.

For the designers deciding whether to obtain a package, not only is the efficiency-generality question important but also the effort required to develop the application versus the effort required to understand and implement the package must be considered. Generally, we overestimate what is required to learn how to use or modify an applications package and underestimate the effort required to develop a similar package ourselves. As a result, most designers would rather develop a system than learn and install a package. There is also the issue of human resistance, as it is probably more of a professional challenge for a programmer to develop a completely new system than to learn how to use and modify a package provided by some outside vendor.

Despite some of the problems described above there has been a trend toward a greater use of packages. One argument against packages is that their generality causes the computer to be used inefficiently. However, hardware has become more efficient and prices per computation have been dropping steadily. Thus the tradeoff between saving machine time and programmer and analyst time has been shifting more in favor of the inefficient use of machines. It may be cheaper to use the computer inefficiently than to pay for a custom-designed system! Packages are also becoming more effective as vendors learn how to design better user interfaces. Not only are the costs of developing systems increasing while the costs of computer systems are dropping, but also the problems of design and management produce uncertainty as to whether an application will ever be implemented. In general, it does require more time to develop a new system than to modify an applications package. If the package will work in the given situation, its use should shorten the development cycle and help the information services department appear more responsive to users. For a more comprehensive discussion of packages, particularly their acquisition, see Fient (1973).

IMPLEMENTATION

Experience has shown that applications packages are not a panacea for all problems, especially large systems like PARS. A programming staff is still required to understand, parameterize, and test the package and train users. On a smaller scale this effort is required for many simple applications packages. A package should be viewed as a way to reduce programming and design time. It is still necessary to undertake systems analysis and preliminary design work with users to see if the package can be used in a given situation. Users must be shown the benefit of changes in their procedures which may be required before they can use the package. A preliminary systems design feasibility study should still be conducted and the user should be presented with various alternatives. These alternatives should include the time, cost, and nature of the changes required in the user's procedures for the alternatives of building a system from scratch versus the acquisition of an applications package.

EXISTING DESIGN AND PROGRAMMING AIDS

TYPES OF AIDS

A number of different programming and systems design aids have been developed with prices ranging from $100 for simple debugging packages to $50,000 or more for special translators (Falor, 1970). There are available several conversion, translation, and simulation aids, for example, to convert a second-generation computer language like Autocoder into Cobol or Basic Assembly Language. Debugging and testing routines are available which generate test data, provide traces of program execution, etc. There are also documentation aids such as flow chart generators and cross-referencing programs to label program variables. Job control language generators are available which allow the user to employ high-level statements to generate detailed job control statements for operating systems. Decision table pre-processors are available to translate tables into Cobol or Fortran. There are a number of shorthand languages or generators on the market; for example, one can use codes for Cobol key words and the appropriate full text will be generated. Other packages are available to manage libraries of source programs and produce variable cross references.

OTHER AIDS

There are a number of modules or packages which can be supplied as a part of the operating system to ease programming tasks. For example, many operating systems provide the ability to catalog and save job control language statements which are executed repeatedly. Another helpful feature in a system is an INCLUDE verb such as the one found on the compatible time-sharing system (CTSS) at MIT. On this early time-sharing system (which has been replaced by the MULTICS system) it was possible to establish a file of source language statements and insert the INCLUDE verb any place in the program. The CTSS system would copy the file of source language statements in the source program at that point. Other useful programming aids are capabilities similar to the "ON" conditions of PL/1. On End of File the user can transfer automatically to some statement in his program. These ON conditions should be extended by language designers; for example, on a certain logical condition the ability to print the values of specified program variables would help in debugging.

There are also available a number of program optimizers which provide execution times and frequencies associated with various parts of a program. These monitors are generally run as part of a software monitoring package and can be useful for tuning a program under development. It is probably not desirable to use them extensively in optimizing an existing program, because changing debugged code for the sake of efficiency is dangerous. However, during the program development cycle the optimizing monitors can help the programmer learn which modules are executed most frequently and what portions of the program are the best candidates for tuning. At least one new computer system provides such a program profile automatically at the request of the programmer.

PROPOSED AND PROTOTYPE AIDS

ISDOS

A system called ISDOS is under development as a research project at the University of Michigan to automate the technical parts of system building (Teichroew and Sayani, 1971). ISDOS stands for Information Systems Design and Optimization System; the research is being conducted in the Industrial and Operations Engineering Department under the general direction of Professor Daniel Teichroew. This research is dedicated to the idea that there is more to the systems development task than simply programming, and that too long a time period elapses between the feasibility study and the implementation of a system. The ISDOS system is being developed to reduce the time and effort required once the basic specifications for the system have been delineated. The development of a powerful package like ISDOS should make it possible to spend more time and effort on organizational behavior problems in systems design.

ISDOS is used after the management requirements for a system have been specified. Its first major component is a problem statement language which the systems designer uses to express processing requirements. He describes what is to be done rather than how; the problem statement language is a concise description of user needs. The problem statement analyzer program analyzes the problem statement language for correct syntax and develops a comprehensive data and function dictionary, checking both static and time-dependent relationships in the data. The output from the statement analyzer is an error-free description of perceived user requirements. Operations research techniques are used to search and select one of a range of alternative systems designs. ISDOS also needs a statement of hardware availability to produce specifications for program modules, storage structures, and scheduling procedures.

The data reorganizer takes the desired storage structures, data definitions from the problem statement analyzer, hardware specifications, and data as they currently exist and stores the data on selected devices in the specified form. The code generator accepts specifications and organizes the problem statements into programs. The system director accepts the code generator, the timing specifications, and the data organization specifications and produces the target information processing system ready to accept inputs from the environment and produce the required outputs.

One portion of the ISDOS system is complete. The Systems Optimization and Design Algorithm (SODA) executes on several computers. In 1971, this program was limited to the design of uniprogrammed batch systems with sequential auxiliary storage organization, linear data structures, and a single CPU. The design algorithm begins with a statement of the processing requirements for an application and generates a complete systems design, including the selection of a CPU and memory configuration. The program incorporates a problem statement language and analyzer along with procedures to select alternative hardware configurations and to generate alternative program and file structure designs. The performance of each alternative is evaluated and optimization techniques are used to select the best alternatives, including network analysis, branch and bound, and nonlinear programming and integer programming algorithms (Nunamaker, 1971).

The ISDOS project is currently under development; it has the potential for becoming a tremendous aid in system building. It should reduce the lead time in developing systems and produce systems which are accurate and work as specified. These two advances will increase the quality of service from the users' standpoint. Also, a great deal of flexibility will be introduced into the design process. If user requirements change, it will be possible to alter the problem statements and regenerate the system for a predictable cost. If the hardware is changed, this system will be recompiled automatically. Too often users have watched their processing bills increase as new hardware is acquired and programs are converted in a computer center while no new output is produced. Instead of converting individual programs, it would be possible to recompile the entire system by using ISDOS. ISDOS should help to alleviate one of the most onerous tasks of systems building, the preparation of programs and files to meet system specifications.

A PROPOSED SYSTEM [1]

The computer has been applied to developing electronic circuits and the design of automobiles, but it has been used very little to help automate systems design itself. The system proposed below is not so revolutionary as the ISDOS project; but it could be developed and implemented more quickly than ISDOS and would be suitable for use in places where ISDOS is not available owing to cost or machine limitations. The proposed prototype system is based on an inexpensive minicomputer devoted to the systems design task. The system used a small central processing unit and a few terminals, plus one or two direct access storage devices. If the system were produced in quantity, it should sell for well under $200,000.

The first component of the system is a heuristic program for systems design to help define system processing stages, given equipment limitations (Lucas, 1971b). The analyst begins at the final output stage and works toward the initial input stage. The system assists the analyst in creating report definitions, master files, intermediate files, and unit record inputs. The program eliminates duplication and checks that all necessary data are maintained throughout the different processing stages. The program is iterative and interactive; the user can back up and take a new path at any point. When the final system is defined, it has

[1] The prototype system discussed in this section has been partially implemented as a research project and is not available commercially.

already been documented, and this documentation can be used by the analyst and programmer to implement the various system modules. A prototype heuristic program is available on the HP 2000C minicomputer time-sharing system at the Graduate School of Business at Stanford (Lucas, 1971b).

A second component of the proposed computer-aided systems design package is a program identifier storage and retrieval system. As systems have developed, programmers have used different variable names or identifiers for each program; as a result the same data are defined by different names in different programs. Frequently these variable names are not adequately documented, and many problems are encountered when a program has to be modified. A prototype information storage and retrieval system for these variable names is also running on the HP 2000C (Lucas, 1973b). The analyst or programmer can inquire on key words used to describe a program variable or identifier when he is developing a program. He can see if the name has already been entered, either as a global variable for the entire system or as a local variable within one program. If a search fails to locate the variable name corresponding to the key words, the programmer defines and enters a variable name for the item. After the system is completed and all programmers have entered their variable names, a set of common variable names will have been used throughout the entire system. Documentation will also have been gathered automatically on-line in the form of a library of variable definitions. For such a system to succeed, programmers will have to be encouraged to use it. Hopefully the documentation aid and maintenance advantages will overcome the initial time requirements to use the system during program development.

The next component of the proposed system is a CRT Report Generating Program which also operates on the HP 2000C using a Tektronics storage tube CRT (Lucas, 1974b). Because of the limited size of the CRT screen, a report is divided into four quadrants. The analyst uses the system to enter a format line, names of variables in the program which will produce the line, and sample data ranges. A teletype terminal is used to enter this information and the appropriate format appears on the screen of the CRT. It is possible to indicate that columns and rows are to be repeated, for example, in printing an array or table. When the entry of information is completed, the analyst can edit the report using the CRT to move lines and portions of lines. The system features the capability to generate pro forma reports containing the sample data values provided earlier by the analyst. When the analyst is satisfied with the output, he can generate PL/1 source language statements with the proper programming variables and format specifications to produce the report.

This component of the computer-aided systems design package accomplishes several objectives. First, it relieves the analyst or programmer of one of the most unpleasant jobs of programming: the development of preliminary report formats and the experimentation required to produce the final report. Second, it provides much flexibility in report design. Instead of giving a

TABLE 7.1. EXAMPLES OF MONITORED DATA

For Each Program Module	For Each Run
Identification	Compile time
Project	Load time
Program	Execute time
Programmer	Program size
	Date
Program Characteristics	Time
Batch or on-line	Completion status
Operating system or application	Programmer data (e.g., type of test)
Computation	
I/O	
Size	
Execution time goal	
Begin date	
Scheduled completion	
Scheduled total computer time	

user a list of the items to be contained in a report, he is shown a pro forma report with sample data, and modifications can be made in the report at low cost. After the system has been developed, it is possible to keep the specifications off-line and reload them for on-line access to make quick and easy modifications as requirements change.

The final component of the proposed dedicated minicomputer system for systems design is a monitoring package to assist with some of the major problems in project management (Lucas, in press). A software monitor is used in the computer where programs are tested; this monitor collects data on the performance of test programs shown in Table 7.1. These data are sent to the minicomputer for analysis and display by programming management. A number of statistics are produced summarizing the results of testing as shown in Table 7.2. The system can be used in daily

TABLE 7.2.   EXAMPLES OF DESCRIPTIVE STATISTICS

Statistics	*Produced for*

Average number of tests
    Compile
    Load
    Execute
	Batch vs. on-line programs
	I/O vs. computational programs
	Operating system vs.
	    application programs
	Differing program sizes
	Each programmer

Average length of time
    Compile
    Load
    Execute	

Average elapsed time by stage
    Compilation
    Test data execution
    Unit testing
    Combined module testing
    Parallel tests	

Average difference between
    Actual completion vs. scheduled
    Actual machine time vs. scheduled	

management to help show the critical development path and assign more resources where needed. It is important that the performance data not be used punitively; programmer cooperation is needed in order to enter the necessary parameters on the job card to start the monitor. The purpose of the system is to help manage the programming effort better, not to threaten programmers.

Not only is the system useful in determining progress for a current project, but also by maintaining historical data on a number of projects, the design staff using the system is building a data base for further analysis. This rich source of data can be used to make future estimates of the time required to develop systems and to obtain a new project. Analysis of these historical data should help forecast how much time will be required for the development of a similar system in the future, historical analyses can be made of programming time, number of tests, amount of computer time required, etc. From a research standpoint it would be possible for programming and systems management to look at the different requirements of batch and on-line systems, operating systems programming versus applications programming, and so on. This particular monitoring system is in the proposal stage; hopefully it will be implemented because it is important that organizations begin to gather data on which to build a better understanding of the programming and systems development process.

All the components described here both in prototype and in proposal form can be combined into a dedicated minicomputer system which is used for aiding the systems design process.[2] See Figure 7.1. A small investment in such a system should lead to significant returns in improving the systems design process. Both the length of time and the amount of effort required to develop and modify an information system should be reduced. The quality of the system should improve, and the responsiveness and flexibil-

[2] It does not appear that a similar system exists or that any organization is planning to develop a system resembling the one proposed here. (Only parts of the proposed system have been implemented to date in a research environment.)

FIGURE 7.1. A PROPOSED SYSTEM FOR COMPUTER-AIDED SYSTEMS DESIGN.

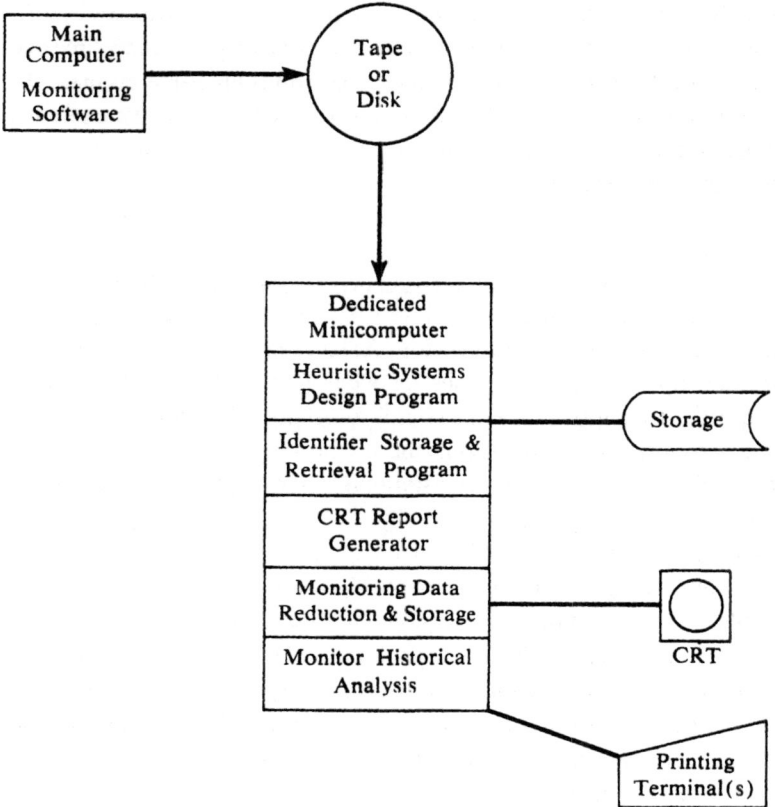

ity of the information services department as perceived by users should increase.

SUMMARY

A number of existing and proposed aids to the systems design process have been discussed in this chapter. It is the responsibility of management to provide cost effective tools to ease the burden of system building. We have maintained that the major components of systems design are basically human and intellectual in nature; mechanical aids can increase productivity and leave more time for attending to other matters, particularly solutions to behavioral problems. Even a system as comprehensive as ISDOS will not solve all of the problems of systems design; organizational behavior issues will still remain along with the problems of project management.

RECOMMENDED READINGS

Falor, K. 1970. "Survey of Program Packages—Programming Aids," *Modern Data,* 3 (8):62–72. (A listing of a number of program packages available to assist in systems design and implementation.)

Lucas, H. C., Jr. 1974. "A CRT Report Generating System," *Communications of the ACM,* 17 (1):47–48.

Teichroew, D. and H. Sayani. 1971. "Automation of System Building," *Datamation,* 17 (16):25–30. (A description and overview of the ISDOS project at the University of Michigan.)

CHAPTER EIGHT

PROJECT MANAGEMENT

INTRODUCTION

IN CHAPTER 2 three major problem areas in systems design were introduced: technical, behavioral, and project management. Without successful project management, creative systems design techniques will be ineffective. For our purposes, project management is concerned with almost the entire systems development cycle including the feasibility study, systems design, the development of specifications, programming, testing, conversion, and implementation. The major difficulties with project management revolve around its uncertainty. There is a lack of standards and knowledge of how to manage computer system projects; they are essentially research and development efforts.

Because the user is unaware of his decision processes, uncertainty for the design team results and project management is affected. This problem was discussed in some detail in Chapter 5, and methods were suggested for working with the user to help him develop a model of his decisions and actions. An example of this type of development effort is Scott Morton's management decision system (see Chapter 6).

The analyst may not understand what the manager needs and may thus add to the level of uncertainty in the design project. The importance of analyst and user working together to reduce this uncertainty has been stressed. There is a need for an iterative approach to design with many reviews by users. Uncertainty also occurs in translating system requirements into programming specifications. In the livestock management system discussed in Chapter 6, a creative approach to reducing this uncertainty was developed in which the users rather than the designers undertook the task of developing program logic and specifications.

Another level of uncertainty is introduced when the programmer tries to interpret programming specifications, develop a program which meets these specifications, and complete the job on schedule. Some of the mechanical aids, like ISDOS, discussed in the previous chapter would help to solve this problem by automating many of these procedures. In the absence of a system like ISDOS, several management strategies for programming are discussed below. New views are emerging which have the potential for revolutionizing the way in which we program computer applications.

Program and system testing are required to reduce uncertainty and validate the systems design. Here again it is useful to have the users participate in the development of the testing program. The livestock management system demonstrates how this can be done; for example, by the creation of fictitious animals for system testing. Through a game, potentially destructive conflict was turned into constructive competition to develop and test the system.

A final uncertainty in design is the degree to which users are ready for and will cooperate with the implementation of the system. In previous chapters we suggested a number of techniques for user involvement; they are particularly important during the implementation stage. Planning for implementation should begin early in the project development cycle. The potential impact on different users and user groups should be forecast (see especially Mumford and Ward, 1968).

PROJECT ORGANIZATION

DESIGN TEAMS

The information services department carries out two separate functions, the design of new systems and the operation of existing systems. The operation of existing systems is a production problem; it is necessary to meet processing schedules while accommodating unusual or unexpected requests. However, the tasks involved are fairly structured and similar to any manufacturing activity in which sequencing is important and schedules must be met.

Systems design, on the other hand, is more closely aligned with research and development. The most popular form of organization for systems design is the project team or matrix organization (De Maagd, 1970). The basic organization chart resembles a matrix; programmers and systems analysts form the rows and projects form the columns. Managers for programmers and systems analysts are employed to solve special problems, conduct salary reviews, recommend promotions, and perform general manpower allocation. At the same time, project managers coordinate teams of analysts and programmers who work on individual projects. A third row should be added to the matrix; in addition to programmers and analysts, users should also be team members (see Table 8.1).

One advantage of the project team organization is that the team becomes a small group within the organization responsible for its own product. Group norms develop and the project team operates with autonomy and self-control. This small, mission-oriented group can be highly flexible in meeting the requirements

TABLE 8.1.   MATRIX ORGANIZATION FOR SYSTEMS DESIGN

Management:	Manager of systems design Manager of analysts * Manager of programming * Project managers $(1, \ldots, n)$			
	Project 1	*Project 2*	*Project 3 . . . Project n*	
Programmers Systems analysts Users				

* Optional depending on the size of the organization.

for systems design. The group, however, may have problems dealing with the ambiguity inherent in design; they will have to define their own tasks and will need help from management to clarify their responsibilities.

The project leader for the team is usually someone from the information services department. However, some organizations are beginning to place a user in overall charge of systems development projects. If a knowledgeable user is available, he is likely to increase user inputs and assure more user participation in design. The use of a noninformation services department employee as a project leader requires a helpful and constructive attitude on the part of the information services design staff. The staff should agree with the selection of the project leader and be willing to cooperate with him.

PROGRAMMING MANAGEMENT

Several early articles on the management of programming projects ask the question whether programmers are so highly specialized that a different form of management is necessary for them (Jones and McLean, 1970). Everyone seems to agree that there are no standards as to what makes a good program or a good programmer. In fact, suggested programming standards are often in conflict with each other. For example, it could be difficult simultaneously to minimize debugging time and to maximize the execution efficiency of a program.

Weinberg (1972b) has provided a great deal of insight into the problem of programmer management; he claims that programmers can obtain clearly stated objectives. Weinberg describes a series of experiments in which program objectives were clearly specified by management for the programmers involved. These included (1) minimum core use, (2) output clarity, (3) program clarity, (4) minimum number of statements, and (5) minimum hours in development. Five teams were involved in writing the programs, and each team was ranked on all five objectives. Each team had its highest rank on the objective it had been given by management! The results of the study also suggest that certain objectives may be incompatible: for example, output clarity and using the minimum number of statements.

Another experiment conducted by Weinberg (1972a) provided programmers with the objectives of either prompt completion of the program or efficient code. The group in which prompt development was emphasized required 29 runs to create a program which was executed in six units of time. The group emphasizing efficient programming required 69 runs to produce a program with a mean execution time of one unit. In most organizations, management does not make objectives clear for programmers. The results of these experiments suggest that programmers can achieve objectives. However, not all objectives can be attained simultaneously, and management has the responsibility to determine what objectives should be stressed for each project. For most commercial information systems development projects, management is probably primarily interested in the objectives of short development times and program clarity to facilitate debugging and future changes to the system.

Establishing clear and compatible objectives for programmers appears to be essential, but it alone is not sufficient to solve the problems of managing the programming and development effort. Weinberg (1972a) has suggested a revolutionary approach to programming called "egoless programming." Until the present time, programming has, in general, been thought of as an individual, creative task. Even though a group may work on different programs, usually a system is so designed that a complete program is assigned to one programmer. Weinberg suggests that programming is really a group activity.

Most programmers have tended to be detached from coworkers, but highly attached to their own programs. If a program proves defective for some reason, it reflects on the programmer, who then becomes highly defensive. We need to restructure the social environment surrounding programming and the attitude and value system held by programmers. In the team approach advocated by Weinberg and Baker, programmers exchange code and examine it; team members study each other's programs and look for errors (Weinberg, 1972a; Baker, 1972a,b). Programmers who practice "egoless programming" expect that there will be errors in their code. They enlist the aid of other members of the team to find errors and in turn cooperate with these other team members to locate their mistakes. Weinberg (1972a) offers many anecdotes describing the benefits of this approach. To generalize from his examples, "egoless programming" drastically reduces the amount of time and expense required to program and debug a system. For evidence of the advantages of the programming method and the use of a programming team supervised by a chief programmer, see Baker (1972a). Another benefit of this approach to programming is that each programmer will understand the entire system better; he will be familiar with programs other than his own.

The major problem in implementing this approach to programming has been the attitude of project management. Managers have tended to believe that programming is an individual task and have often broken up teams that work too closely together. Management should create an environment conducive to team programming; a reward structure can be established to encourage teamwork. With respect to the assignment and staffing of project teams, it may be desirable to treat programming teams as units rather than to assign individuals to each project. Certainly, the programmers on a team should be involved in work assignment decisions.

Even with a programming team which is working well, good management can contribute to progress. First, managers should encourage honest progress reports by adding resources rather than punishing those who offer realistic, but possibly discouraging, reports. Owing to the nature of programming and the norms that will be developed by the project team, management should be very flexible and tolerant of team behavior. For example, the teams may desire to work strange hours and management action may be required to make this possible. In essence, management has to provide the team the freedom it needs. For a creative group of computer programmers an atmosphere of freedom and somewhat permissive leadership should result in the best performance (Mumford and Ward, 1968).

A project manager cannot be expected to know individual programs as well as does the programming team; the job of the manager is to manage, not to program. Unfortunately, some program managers act as if they were experts about individual programs and abstruse technical problems; such managers quickly lose the respect of their staff. The technical staff is available to advise the manager and should be enlisted to help make decisions on technical problems.

TECHNICAL GUIDELINES FOR PROJECT MANAGEMENT

GENERAL

Special techniques have been suggested to improve the management of software projects (Kay, 1969; Jones and McLean, 1970). First, it is important to set objectives at the beginning of the project. Projects have also failed because management has not assessed the possible; that is, what is the state of the art for the staff in the organization? Past experience indicates that some type of project plan, such as a critical path chart, is necessary. Methods should also be developed to encourage documentation, for example, by providing on-line systems like the project management system proposed in the previous chapter. It is also important to use a system librarian to control procedures for making changes to specifications.

In addition to some of these general suggestions, there are research findings available to offer technical guidelines for project management. The primary focus of these suggestions is on planning; a system which is planned will take less overall time to develop, even though effort has to be expended in the early stages to develop plans. Programmers are always eager to begin programming; however, the more planning and the better the specifications, the fewer will be the changes and the shorter will be the overall development time.

HIGHER LEVEL LANGUAGES

The first specific recommendation that can be made for most information systems development projects is that a higher level language be used. There was much debate about this issue a number of years ago, but it has been resolved in favor of higher level languages. In fact, higher level languages have even been used in systems programming, and at least one manufacturer has designed his hardware to provide a natural environment for higher level language execution (Corbato, 1969). A higher level language facilitates communications among programmers. Though not fully self-documenting, higher level language code is more self-documenting than assembly language. A higher level language is also more highly compatible across systems; there is more flexibility in making changes to the system and program maintenance is easier. Generally, one expects to find a smaller number of programming errors when a higher level language is used. In the system described by Corbato (1969) most errors were found in mismatched declarations or inverted parameters in calling sequences.

MODULAR PROGRAMMING

Another technique being used more and more to reduce programming time and cost is modular programming. In its simplest form, modular programming simply means breaking a system into many parts. Each part is easier to understand and debug, and it is easier to change the system when each function is a separate module. Modular programming requires more effort in design and more management to coordinate the parts, but it should speed programming and debugging. Separate groups can also work on different modules independently of each other, except for interfacing requirements.

Recently, a criterion for dividing systems into modules was suggested to simplify modifications to the system (Parnas, 1972). A commonly used criterion has been to make each major step a module; that is, each major box on a flow chart becomes a separate module. Parnas has suggested another criterion, that of "information hiding." In this scheme, modules do not correspond to steps in processing; each module contains a unique design decision. In order to assign modules, one begins with a list of difficult design decisions and then develops modules which hide these decisions from other modules. The interface between modules is designed to reveal as little as possible about each module's inner workings.

Parnas compares two examples of modularization for a single problem to demonstrate how his suggested criterion makes changes much easier. The programming problem is the development of a Keyword in Context (KWIC) index which permutes the titles of documents and alphabetizes the permuted titles on the keywords. One major design decision in this system is to keep all the lines of text to be processed in core storage. For the first example of modularization, the input routine is one module which reads data lines from the input medium and stores them in core for processing by other modules. All remaining modules assume that data lines are in core.

The second example of modularization requires two modules. The first is an input routine which reads data lines from the input medium and calls the second module, line storage, to store the lines internally. The line storage module provides the user with the option to retrieve any characters of any line.

If a design change is necessary because it is suddenly recognized that there may not be enough room for all lines in core, the "information hiding" modularization in the second example is easier to change; it has isolated the design decision to store all lines in core from other modules. In the first example, all modules would have to be altered since they all assume that the lines are in core. In the second example, only the line storage module has to be altered to reflect the modified design decision.

Following the "information hiding" approach, each module may not be a separate procedure with associated calls owing to efficiency considerations. Instead, a module may be a portion of code inserted into a program; routines and procedures are assembled from a collection of code from different modules.

STRUCTURED PROGRAMMING

Another technique which can reduce time and cost in developing systems is structured programming. This idea was first suggested by Dijkstra (1968) and has been applied to information systems development by Baker (1972a,b). Structured programming represents a set of rules that enhance a program's readability and ease of maintenance. It is based on a theorem in computer science which states that any program with one entry and one exit can be written using the following constructs:

(1)   sequential operations (including Procedure Calls)
(2)   IF THEN ELSE
(3)   DO WHILE

Using only these three statement types is somewhat restrictive and forces more thought in programming. However, with the elimination of GO TO's it is possible to read a program from top to bottom with no intervening jumps. The programmer can easily see the conditions required for modifying a block of code. Tests are easier to specify and documentation is enhanced. Structured programming should be combined with code indentation and formatting to enhance the readibility of the program. Comments are also useful, particularly where procedure calls are executed.

TOP-DOWN PROGRAMMING

Baker (1972a) has advocated the use of another technique with structured programming called "top-down" programming. Most systems are designed from the top level down, but are implemented from the bottom up; that is, the basic modules are written first and then integrated into subsystems. Baker advocates reversing this process by writing the highest level programs first and testing them while the next lower level is being written. Dummy subroutine and procedure calls are used for lower level modules which have not yet been written. Top-down programming makes the status of work clearly visible and shows what functions must be performed by lower level routines. Interfaces are defined before the development and coding of functions using the interface. Baker (1972a) reports that the use of group programming managed by a chief programmer, structured programming, and the top-down approach resulted in a successful project which was delivered and accepted on time. Though this represents only one case, there is strong evidence that a combination of these specific techniques for managing software design can drastically improve systems development.

SUMMARY

Creative systems design can succeed only if the design project is well managed; however, there is no one approach which assures success in project management. A variety of guidelines to improve project management are summarized in Table 8.2. The most important activities are establishing goals, planning, developing an organizational structure, and comparing project status with the plan. The treatment of programming as an "egoless" group activity is essential. In order for the project to be managed, realistic inputs from programmers are needed. Management must create an environment which focuses on building a better system on schedule; realistic estimates can be encouraged by providing additional resources for the programming team.

Management controls the project; it monitors progress, provides rewards, and reallocates resources to assist groups which have encountered difficulties and delays. This planned approach to project management, combined with good technical design practices including the use of modular programming, structured programming, top-down programming, and a systems librarian, should increase the probability for successful project management. The final result ought to be the development of useful systems with minimum time and cost overruns.

TABLE 8.2.   SUMMARY OF PROJECT MANAGEMENT TECHNIQUES

Design team organization	Technical management
Matrix organization	General
Capable project leader	Assess state of the art
	Establish project plan
	Encourage documentation
	Establish system librarian
Programming management	Higher level languages
Establish objectives	Modular programming
"Egoless" programming	Structured programming
Programming teams	Top-down implementation

RECOMMENDED READINGS

Baker, F. T. 1972. "Chief Programmer Team Management of Production Programming," *IBM Systems Journal*, 11 (1):56–73. (An example is presented of systems design using structured programming, chief programming teams, and "egoless" programming.)

Corbato, F. J. 1969. "PL/1 as a Tool for System Programming," *Datamation*, 15 (5):68–76. (A discussion of the advantages of using a higher level language even for systems programming.)

Jones, M. M. and E. R. McLean. 1970. "Management Problems in Large Scale Software Development Projects," *Sloan Management Review*, 11 (3):1–16. (A discussion of techniques which can be used to improve the management of software development projects.)

Kay, R. H. 1969. "The Management and Organization of Large Scale Software Development Projects," AFIPS Conference Proceedings (SJCC), 34, Montvale, New Jersey: AFIPS Press: 425–33. (A good article describing the reasons for the failure of a number of software projects; suggestions are made for improving the management of these projects.)

Weinberg, G. M. 1972. *The Psychology of Computer Programming.* New York: Van Nostrand Reinhold. (A stimulating book which offers many insights into the programming process; studies are presented in an easy-to-read, nontechnical manner and much anecdotal evidence is presented to support the author's views.)

———. 1972. "The Psychology of Improved Programming Performance," *Datamation*, 18 (11):82–85. (The results of a study are presented in which programmer teams were given a series of objectives and programs to write; each programming team ranked first on its primary objective.)

CHAPTER NINE

IN CONCLUSION

CREATIVE SYSTEMS DESIGN TECHNIQUES are intended to solve the organizational behavior problems associated with the development of information systems. Designers have to diagnose user needs and work collaboratively with users to solve them. There are a variety of ways to encourage cooperation if top management, users, and the information services department have created the proper environment.

One of the central themes of our philosophy is that *the systems designer is a catalyst for the user who designs his own system.* The systems analyst helps the user to understand his needs and builds a descriptive model of current information processing and decision procedures. The model is refined with the user's assistance and a new information system is developed if feasible and desirable. The analyst assists the user in converting his needs into an information system design and in translating the design into a computer system.

Creative design means considering the physical interface between the users and the system. The designer should place himself in the position of the user and work collaboratively with him when developing manual procedures and input/output requirements. Our philosophy means that we define and evaluate the quality of a system from the user's perspective independently of technological considerations.

Though creative systems design is primarily related to organizational behavior problems, technical issues and project management are vital to the success of a system. The use of creative techniques will be fruitful only if all technical components succeed and the project is managed successfully. For these reasons, we have discussed mechanical aids to design and project management.

If the approaches to systems design suggested here are followed, what are the likely results? As the reader may have concluded by now, creative systems design practices require considerable effort. The short-run dollar costs of an analyst or technician designing and implementing a system he thinks meets user requirements are far less than the costs of creative design. To follow the recommended approaches, more time will be required from users, the systems design staff, and management. Each system will cost more and take longer to develop. However, the major argument in favor of this approach is the extreme dissatisfaction and failure of so many systems which have been developed under traditional non-user-oriented approaches. Designing a few high quality systems which contribute to the objectives of the organization and are well received is vastly preferable to developing a large number of systems which are unused or dysfunctional.

The results of practicing creative systems design will probably be a reduction in the number of systems developed in the organization. However, the systems which are implemented should accomplish more and should be used. The greater time and effort required to plan and develop a system should be offset by the quality and acceptance of the systems which are implemented. The widespread user dissatisfaction with current systems will be strengthened as more complex and sophisticated systems are developed unless we learn how to practice creative systems design. A change in design emphasis is being suggested from quantity and speed to quality and effectiveness.

Computer-based information systems have a tremendous potential for assisting management and the organization. The challenge is to develop systems which will allow us to realize this potential. Acceptance of the user-oriented philosophy of creative systems design and the implementation of the creative approaches to systems design described in this monograph should help to meet this challenge.

REFERENCES

Ackoff, R. L. 1967. "Management Misinformation Systems," *Management Science*, **14** (4):B147–56.

Anthony, R. 1965. *Planning and Control Systems: A Framework for Analysis*. Boston: Division of Research, Graduate School of Business Administration, Harvard University.

Baker, F. T. 1972a. "Chief Programmer Team Management of Production Programming," *IBM Systems Journal*, **11** (1):56–73.

———. 1972b. "System Quality through Structured Programming," *AFIPS Conference Proceedings* (FJCC), 41, Montvale, New Jersey, AFIPS Press: 339–43.

Buxton, J. N. and B. Randell. 1969. *Software Engineering Techniques*. Rome: NATO Science Committee.

Byrnes, C. J. and D. B. Steig. 1969. "File Management Systems: A Current Summary," *Datamation*, **15** (4):138–42.

Churchill, N. C., J. H. Kempster, and M. Uretsky. 1969. *Computer-Based Information Systems for Management: A Survey*. New York: National Association of Accountants.

Corbato, F. J. 1969. "PL/1 as a Tool for System Programming," *Datamation*, **15** (5):68–76.

De Maagd, G. R. 1970. "Matrix Management," *Datamation*, **16** (13):46–49.

Dijkstra, E. W. 1968. "The Structure of 'THE' Multiprogramming System," *Communications of the ACM*, **11** (5):341–46.

Driver, M. J. and S. Streufert. 1969. "Integrative Complexity: An Approach to Individuals and Groups as Information-Processing Systems," *Administrative Science Quarterly*, **14** (2):272–85.

Falor, K. 1970. "Survey of Program Packages—Programming Aids," *Modern Data*, **3** (8):62–72.

Fient, H. G. 1973. "Management of the Acquisition Process for Software Products," *Management Informatics*, **2** (3):153–64.

Gerrity, T. 1971. "Design of Man-Machine Decision Systems: An Application to Portfolio Management," *Sloan Management Review*, **12** (2):59–75.

Gorry, G. A. and M. S. Scott Morton. 1971. "A Framework for Management Information Systems," *Sloan Management Review*, **13** (1):55–70.

Harper, W. L. 1973. *Data Processing Documentation: Standards, Procedures and Applications.* Englewood Cliffs, New Jersey: Prentice-Hall.

Hickson, P. J., C. R. Hennings, C. A. Lee, R. E. Schneck, and J. M. Pennings. 1971. "A Strategic Contingencies Theory of Interorganizational Power," *Administrative Science Quarterly*, **16** (2):216–29.

Jones, M. M. and E. R. McLean. 1970. "Management Problems in Large-Scale Software Development Projects," *Sloan Management Review*, **11** (3):1–16.

Katz, D. and R. K. Kahn. 1966. *The Social Psychology of Organizations.* New York: John Wiley and Sons.

Kay, R. H. 1969. "The Management and Organization of Large Scale Software Development Projects," *AFIPS Conference Proceedings* (SJCC), 34, Montvale, New Jersey, AFIPS Press: 425–33.

Lawler, E. E. and R. J. Hackman. 1969. "Impact of Employee Participation in the Development of Pay Incentive Plans: A Field Experiment," *Journal of Applied Psychology*, **53** (6):467–71.

Lucas, H. C., Jr. 1971a. "A User-Oriented Approach to Systems Design," *Proceedings of the 1971 ACM Annual Conference:* 325–38.

———. 1971b. "Computer Aided Systems Design: Systems Analysis Program," Stanford: Graduate School of Business Technical Report No. 6.

——— and R. B. Plimpton. 1972. "Technological Consulting in a Grass Roots, Action Oriented Organization," *Sloan Management Review*, **14** (1):17–36.

———. 1973a. *Computer Based Information Systems in Organizations.* Palo Alto: Science Research Associates.

———. 1973b. "Computer Aided Systems Design: Identifier Storage and Retrieval Program," Stanford: Graduate School of Business Technical Report No. 19.

———, K. W. Clowes, and R. B. Kaplan. 1973c. "Frameworks for Information Systems," Stanford: Graduate School of Business Research Paper No. 142.

———. 1973d. "The Problems and Politics of Change: Power, Conflict and the Information Services Subunit," in F. Gruenberger (ed.), *Effective vs. Efficient Computing.* Englewood Cliffs, New Jersey: Prentice-Hall.

———. 1974a. "An Empirical Study of a Framework for Information Systems," *Decision Sciences*, **5** (1):102–14.

———. 1974b. "A CRT Report Generating System," *Communications of the ACM*, **17** (1):47–48.

——, "Performance Evaluation and Project Management," San Diego: *Command and Control Software for the 1980's* (in press).

McCracken, D. 1971. ". . . but the ambivalence lingers," *Datamation* **17** (1):29–30.

McKenney, J. G. 1972. "Human Information Processing Systems," Harvard Graduate School of Business Working Paper No. HBS 72–4.

Mason, R. D. and I. I. Mitroff. 1973. "A Program for Research on Management Information Systems," *Management Science,* **19** (5):475–87.

Mumford, E. and O. Banks. 1967. *The Computer and the Clerk.* London: Routledge & Kegan Paul.

—— and T. B. Ward. 1968. *Computers: Planning for People.* London: B. T. Batsford.

Nie, N. H., D. H. Bent, and C. H. Hall. 1970. *Statistical Package for the Social Sciences.* New York: McGraw-Hill Book Company.

Nunamaker, J. F., Jr. 1971. "A Methodology for the Design and Optimization of Information Processing Systems," Purdue University: Krannert Graduate School of Industrial Administration Paper No. 301.

Parnas, D. L. 1972. "On the Criteria to be Used in Decomposing Systems into Modules," *Communications of the ACM,* **15** (12):1053–58.

Pounds, W. F. 1969. "The Process of Problem Finding," *Industrial Management Review,* **11** (1):1–20.

Proshansky, H. and B. Seidenberg (eds). 1965. *Basic Studies in Social Psychology.* New York: Rinehart and Winston.

Scheflen, K. C., E. E. Lawler, and R. J. Hackman. 1971. "Long Term Impact of Employee Participation in the Development of Pay Incentive Plans: A Field Experiment Revisited," *Journal of Applied Psychology,* **55** (3):182–86.

Schein, E. 1965. *Organizational Psychology,* Englewood Cliffs, New Jersey: Prentice-Hall.

Scott Morton, M. S. 1967. "Interactive Visual Display Systems and Management Problem Solving," *Industrial Management Review,* **9** (1):69–81.

——. 1971. *Management Decision Systems.* Boston: Division of Research, Graduate School of Business Administration, Harvard University.

Sibley, E. H. and A. G. Merten. 1973. "Implementation of a Generalized Data Base Management System within an Organization," *Management Informatics,* **2** (1):21–31.

Simon, H. 1965. *The Shape of Automation for Men and Management.* New York: Harper and Row.

Tannenbaum, A. S. 1966. *Social Psychology of the Work Organization.* Belmont, California: Wadsworth Publishing Company.

Teichroew, D. and H. Sayani. 1971. "Automation of System Building," *Datamation,* **17** (16):25–30.

Trist, E. L. 1963. *Organizational Choices.* London: Tavistock Publications.

Walton, R. E. and J. M. Dutton. 1969. "The Management of Interdepartmental Conflict: A Model and Review," *Administrative Science Quarterly*, **14** (1):73–84.

Weinberg, G. M. 1972a. *The Psychology of Computer Programming*. New York: Van Nostrand-Reinhold.

——. 1972b. "The Psychology of Improved Programming Performance," *Datamation*, **18** (11):82–85.

Whisler, T. L. 1970. *Information Technology and Organizational Change*. Belmont, California: Wadsworth Publishing Company.

Withington, F. G. 1970. "Cosmetic Programming," *Datamation*, **16** (3):91–95.

INDEX